4/12

E4—00

PRENTICE-HALL BIOLOGICAL SCIENCE SERIES

William D. McElroy and Carl P. Swanson, *Editors*

Concepts of Modern Biology Series

Animal Parasitism, CLARK P. READ

Behavioral Aspects of Ecology, 2nd ed., PETER H. KLOPFER

Concepts of Ecology, EDWARD J. KORMONDY

Critical Variables in Differentiation, BARBARA E. WRIGHT

Ecological Development in Polar Regions: A Study in Evolution,
    MAXWELL J. DUNBAR

Euglenoid Flagellates, GORDON F. LEEDALE

Genetic Load: Its Biological and Conceptual Aspects, BRUCE WALLACE

An Introduction to the Structure of Biological Molecules,
    JAMES M. BARRY AND E. M. BARRY

The Organism as an Adaptive Control System, JOHN M. REINER

Processes of Organic Evolution, G. LEDYARD STEBBINS

Critical Variables
in
Differentiation

CONCEPTS OF MODERN BIOLOGY SERIES

William D. McElroy and Carl P. Swanson, *Editors*

Critical Variables
in
Differentiation

BARBARA E. WRIGHT

Department of Developmental Biology
Boston Biomedical Research Institute

PRENTICE-HALL, INC., *Englewood Cliffs, New Jersey*

Library of Congress Cataloging in Publication Data

WRIGHT, BARBARA EVELYN, 1926–
    Critical variables in differentiation.

    (Concepts of modern biology series) (Prentice-Hall biological science series)
    Includes bibliographies.
    1. Cell differentiation. 2. Cell metabolism. 3. Cytochemistry. I. Title.
QH607.W74   574.8'761   72–5631
ISBN 0–13–194209–3

Printed in the United States of America

PRENTICE-HALL INTERNATIONAL, INC., *London*
PRENTICE-HALL OF AUSTRALIA, PTY. LTD., *Sydney*
PRENTICE-HALL OF CANADA, LTD., *Toronto*
PRENTICE-HALL OF INDIA PRIVATE LIMITED, *New Delhi*
PRENTICE-HALL OF JAPAN, INC., *Tokyo*

To Kees

Contents

Metabolic flux control of reaction rates. Metabolic flux control of enzyme levels. Metabolic flux control of enzyme activity by allosteric modulation and end-product inhibition. Metabolic flux control of the initiation, duration, termination, and stability of differentiation.

Life cycle and general metabolism. The role of metabolites. The mechanism of changes in metabolite concentration. The role of enzymes. The mechanism of changes in enzyme activity *in vitro* during differentiation.

Foreword

Concepts of Modern Biology Series   The main body of biological literature consists of the research paper, the review article or book, the textbook,
and the reference book, all of which are too limited in scope by circumstances other than those dictated by the subject matter or the author.
Unlike their usual predecessors, the books in this series, CONCEPTS OF
MODERN BIOLOGY, are exceptional in their obvious freedom from such
artificial limitations as are often imposed by course demands and subject
restrictions.

Today the gulf of ignorance is widening, not because of a diminished
capacity for learning, but because of the quantity of information being
unearthed, most of which comes in small, analytical bits, undigested and
unrelated. The role of the synthesizer, therefore, increases in importance,
for it is he who must take giant steps, and carry us along with him; he must
go beyond his individual observations and conclusions, to assess his work
and that of others in a broader context and with fresh insights. Hopefully,
the CONCEPTS OF MODERN BIOLOGY SERIES provides the opportunity for
decreasing the gulf of ignorance by increasing the quantity of information
and quality of presentation. As editors of the Prentice-Hall Biological
Science Series, we are convinced that such volumes occupy an important
place in the education of the practicing and prospective teacher and
investigator.

WILLIAM D. MC ELROY

CARL P. SWANSON

Preface

This monograph represents an unusual point of view about biochemical mechanisms underlying differentiation. It is not intended to cover the subject, or even a particular aspect of it, but rather to point out that the current emphasis on the control of differentiation at the nucleic acid level may in part be based upon unjustified assumptions, and to draw attention to some important facets of the problem which are currently being neglected. At the present time, the application of our knowledge of intermediary metabolism and steady state kinetics to differentiating systems may be particularly rewarding. This volume will have served its purpose if it stimulates just a few students to proceed in this direction: to stop searching for triggers and ultimate causes; to start a detailed analysis of the *relationships* between various cellular components, *all* of which are essential to the activity of metabolic pathways underlying differentiation; to thus approach, conceptually and experimentally, an understanding of the *critical variables* controlling this metabolism under the conditions of the living cell.

I would like to acknowledge the collaboration of Dr. William Simon and Mr. Timothy Walsh during the initial stages of developing the kinetic modeling approach. In subsequent struggles to become my own programmer, I was aided by Dr. Rosemarie Marshall and Dr. John Gergely. More recently, Mr. David Park has joined our group. He has not only developed a computer program uniquely suited to our analytical requirements, but has contributed significantly to the further sophistication and presentation of this modeling technique. For criticisms of various sections of the manuscript, I am grateful to many colleagues, including Dr. Dave Coleman and Dr. Kathleen Killick;

Dr. Gary Gustafson advised us in the area of enzyme kinetics. Finally, the investigations summarized in this volume were supported by National Institutes of Health Research Grant number HD-04667 COM from the National Institute of Child Health and Human Development.

BARBARA E. WRIGHT

Critical Variables
in
Differentiation

Introduction:
Semantics and a Little Philosophy

> "... the term 'cause-effect' represents a two-term relation, and as
> such, is a primitive generalization never to be found in this world,
> as all events are serially related in a most complex way, independent
> of our way of speaking about them. If we expand our two-term
> relation 'cause-effect' into a series, we pass from the inferential
> level to the descriptive level, and so can apply a behavioristic,
> functional, actional language of order."
>
> *Korzybski, 1958* [2]

As any student of semantics well knows, definitions can be neither true nor false, as they are an arbitrary matter of agreeing about the meaning of words. Definitions can be good or bad, however, and a good one is usually clear, useful, self-consistent, and in accordance with convention. It is my observation that a serious confusion surrounds the use of certain key words in discussions about the mechanism of differentiation. The purpose of this introduction is to formulate and defend my definitions of these words, as a framework for understanding the discussions that follow.

Unless otherwise specified, the term *differentiation* will be used in a broad morphological sense to describe the entire complex of processes resulting, for example, in the transformation of a fertilized egg into a fly or the transformation of myxamoebae into fruiting bodies of the cellular slime mold. When the term is used in a more restricted sense, it will be so indicated and describe the synthesis of a specific component (e.g., cell wall polysaccharide) among the many comprising the complex of processes referred to above. The term *endogenous* as applied to differentiation will be used in a somewhat unconventional manner. It will refer to any material within the differentiating system as a whole, as contrasted to *exogenous* nutrients. Endogenous reserves, therefore, may include various materials within the differentiating system, ranging from the protein within individual cells of the cellular slime mold to the endosperm of seeds or the yolk of eggs.

Any differentiation process—whether it be a complex transformation such as gastrula formation in the sea urchin egg or a relatively simple change such as cell wall accumulation in the cellular slime mold—is dependent upon all of the cellular components present before differentiation occurs. All of the biochemical machinery required for the maintenance of any viable cell must be in operation before, and during, the relatively few additional changes required for a particular cell specialization. The previous specific chemical composition of the cell or cells in question plays a determining role, as does the unique array of genes, messengers, and enzymes already present. Equally critical is the availability of an energy source and of precursor materials used in the synthesis of the end products characteristic of the differentiated cell. The interaction of all these components uniquely describes and distinguishes one organism from another. Each system capable of differentiation evolved *as a unit*, and the details of the transformations that will occur are determined by the unit as a whole. There is no *a priori* reason for singling out any particular biochemical event on which differentiation depends as being more essential or necessary than others. Similar viewpoints have been expressed by others [4].

Figure 1 illustrates an example of a differentiation process: eye pigment accumulation in the fly at time t during the transformation of the fertilized egg (shown at top) to the fully differentiated fly. Pigment accumulation at time t depends upon the participation of all the cellular components shown—and many more. Genes, enzymes, precursors, and energy sources are equally important and necessary: an enzyme has no significance without a substrate, and vice versa; a gene cannot be activated without a stimulant of some kind (e.g., an inducer), which in turn must have a source, and so on, in a never-ending series of events back in time to the origin of the egg, and indeed, into the evolution of this egg-fly system. In a very significant sense, the terms "cause" and "trigger" deny this history of successive events. These words usually describe the interaction of two potentially isolatable systems: for example, "the bacteria caused the man's illness" or "the sperm triggered differentiation of the *Drosophila* egg." The sperm and egg are separable entities with unique histories, and may be observed and described independently of each other. Their interaction is therefore possible, and use of the word "cause" in such contexts is clear and desirable. Turning now to a system

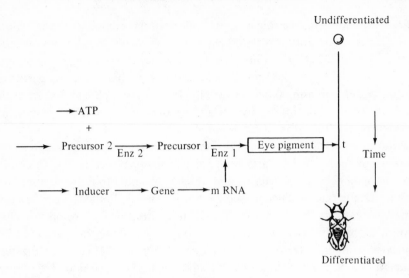

Figure 1.

undergoing differentiation, it is easy to see that no cellular component has inherent causal significance; no gene or metabolite possesses the intrinsic ability to act independently—each has significance only within the framework of its particular cellular environment. Application of the words "cause" or "trigger" to particular components within a differentiating system may delude the user into feeling that there exist certain "spontaneous" or independent factors, uniquely capable of "causing" differentiation to occur. Undue significance may be attributed to an event that is only one of an inevitable series, *all* of which are completely dependent upon previous events and hence are of comparable importance with respect to the particular aspect of differentiation in question. In such a complex system, to abstract one event (e.g., the appearance of a substrate or enzyme) by labeling it "cause" is merely an expression of bias or an indication that a particular event in time is more easily observed or measured than the others: A description of preceding events occurring at specific *points in time*, relative to the event in question, should be substituted for the concept and use of "cause" [5]. To look back too far into the history of an organism for "causes" of a particular aspect of its differentiation is to lose sight of the problem initially posed. Logically, such a search would extend to the evolution of the organism. For these reasons, the term "cause" will not be used to refer to events within a single differentiating system.

The preceding discussion is not simply a superficial exercise in semantics.

4 Introduction: Semantics and a Little Philosophy

Preoccupation with the role of a particular cellular component as a "cause" of differentiation must direct and channel one's investigations (and interpretations) prematurely, and thus curtail an objective unravelling of the unique intracellular milieu and biosynthetic pathways characterizing particular types of differentiating tissue. Hopefully, this monograph will document the validity of this conclusion, by elucidating the role of cellular components not currently studied in differentiating systems.

Since the term *control* as applied to differentiation is often used in a confusing manner, a brief clarification of its meaning here is also desirable: A *controlling* event will refer to one which is essential to differentiation, but by which the rate of differentiation is not necessarily limited—i.e., the event may or may not be limiting for the differentiation process being analyzed. Many events clearly control differentiation; very few thus far have been implicated as limiting factors. Differentiation is a process of cell specialization superimposed upon and dependent upon many nonspecialized functions of the cell. Thus, interference with the maintenance of a messenger RNA (mRNA) for an enzyme essential to energy metabolism may very well stop differentiation. If so, the activity of this enzyme is a prerequisite for, but not necessarily unique to, or limiting, the process of differentiation. In the sense that all of the enzymes necessary for differentiation ultimately owe their existence to genes, the latter can be said to control differentiation. This conclusion is neither profound nor useful, of course, as it could as well be applied to any other cell component that is also essential to differentiation. In order to establish priorities regarding those areas of biochemical research most likely to yield information on mechanism, the problem should be defined more precisely—i.e., in terms of limiting events.

If we are interested in the *mechanism* of differentiation, we wish to determine which cellular components are not only essential, but are also *critical variables*—i.e., *limit the rate of differentiation at particular points in time.* In order to ask meaningful questions about which variables are critical, it is first necessary to define (arbitrarily) a specific differentiation process with respect to (1) the nature of the transformation—e.g., eye pigment accumulation at time *t* in Fig. 1, blastula formation in a fertilized sea urchin egg, cell wall construction in the cellular slime mold, etc., and (2) the time period in the history of the organism (it would be impractical to include its evolution, relevant as that may be). Only then will it be possible to seek meaningful answers. Returning again to Fig. 1, which cellular component might be an immediate or primary critical variable for pigment accumulation from the time the egg is fertilized until time *t*? Perhaps it is gene activation which limits the synthesis of the mRNA and hence the enzyme catalyzing the conversion of precursor 1 to eye pigment. If so, it follows that all of the machinery and precursor materials required for the synthesis of the mRNA, enzyme 1, and precursor 1 are not only present, but in excess—e.g., *not* critical variables. It also follows that inducer availability is an earlier (i.e., secondary) critical variable in this series of events. In fact, for β galactosidase in *Escherichia coli*, the availability of both cyclic AMP and inducer are critical to the initiation of enzyme synthesis [1]. Perhaps the rate of pigment accumulation is not limited by the absence of a catalyst, but by the availability of an energy source or of precursor 1, which was formed at a limiting rate or used in a competing reaction prior to time *t*. After all, the living cell is not a test tube into which we usually put an excess of substrate and limiting levels of enzyme to facilitate the measurement of a reaction rate. On the contrary, enzyme levels *in vivo* are relatively high, and metabolite availability generally limits the rate of a reaction.

Now suppose the gene indicated in Fig. 1 becomes mutated, resulting in a white-eyed fly: Have we learned anything about the mechanism of differentiation? No—only that genes are essential to make enzymes. If, in the normal fly, the critical variable is the unmasking of mRNA or the availability of precursor 1, then the gene is mechanistically irrelevant—one of many essential components, of course, but not a *critical variable*. Types and levels of mechanism may easily be confused. The mechanism of primary interest here is not that of gene activation or enzyme synthesis, but rather, the mechanism of pigment accumulation. We must also try to assess the extent to which one cellular component is limiting in comparison to the others and to determine if more than one factor is limiting. An interesting example to illustrate the complexity of the relationships between interdependent cellular components in their effect on development can be found in the case of cartilage formation; the metabolic machinery is present but unexpressed, except under particular environmental circumstances [3]. Embryonic tissues possess all the enzymes and metabolic intermediates necessary for forming cartilage. This end product only accumulates after induction, however, due to the stabilization or enhancement of a preexisting pattern—not because of the acquisition of a *new* metabolic pattern.

So little is known about the fundamental biochemical mechanisms underlying differentiation that we should at this point, perhaps, be less concerned about the mechanisms by which enzyme levels change and more concerned about the consequences of such changes to the control of differentiation. In the differentiating cell, is the level of enzyme or substrate limiting the rate of an essential reaction? What is the significance and consequence of enzyme accumulation during differentiation: Is such an accumulation critical to, or does it reflect, increased metabolic activity *in vivo*? Perhaps our wide-spread preoccupation with enzymes as the ultimate product of genetic activity has suppressed such simple questions, which may be fundamental to our understanding of the biochemical mechanisms underlying differentiation. Considering our lack of knowledge concerning the biochemical basis of differentiation, it is desirable to be as thorough and as impartial as possible in describing the sequence of events in order to define those that may be critical mechanistically. As we have seen, the *requirement* for a particular cellular component (e.g., gene, enzyme, or substrate) at a specific time during differentiation implies nothing with respect to whether it (1) existed (unused) long before, or (2) limits the rate of the transformation in question.

The suggestion has been made in this introduction that efforts to interpret the behavior of all cellular components involved in differentiation in terms of a single component may interfere with our understanding of the system. Such an approach renders an objective analysis difficult and stresses the role of components which may or may not be limiting the rate of the differentiation process in question. It appears to be no more justified or useful to choose selective gene activation as "the" basis of differentiation than to choose, for example, substrate availability. After all, metabolites are not only essential to gene activation [1], but modify enzyme activity; as substrates, their rate of availability must determine the rate at which the end products of differentiation accumulate. In order to illustrate these conclusions, we will break into the (unbreakable) cycle of interdependent cellular events at a new point for a change and explore the consequences of metabolite availability on the process of differentiation. Thus from a naive, hierarchical point of view, other levels of control would be "post-metabolite," in contrast to the usual (and equally unrealistic) concept of "post-transcriptional" control.

During differentiation in primitive systems, changes in endogenous metabolism are responsible for alterations in the small molecule population which can control the activity of metabolic pathways essential to differentiation by a number of different mechanisms. For example, changes in the rate of availability, or flux, of endogenous metabolites could affect: (1) the rate of substrate-limited reactions critical to differentiation; (2) the concentration of enzymes, by changing their rate of synthesis (induction) or degradation (stabilization); (3) enzyme activity, by allosteric modulation and end product inhibition; and (4) the duration and stability of morphogenesis. The following chapter will attempt to document the importance of changes in metabolite flux in controlling the activity of metabolic pathways by these mechanisms during differentiation.

REFERENCES

1. ERON, L., ARDITTI, R., ZUBAY, G., CONNAWAY, S., and BECKWITH, J. R. (1971). *Proc. Nat. Acad. Sci. U. S.* **68,** 215.

2. KORZYBSKI, A. (1958). *Science and Sanity,* Fourth Edition. Clinton, Mass: The Colonial Press, Inc., p. 217.

3. LASH, J. W. (1968). In "Symposium on Molecular Aspects of Differentiation." *J. Cell Physiol.* **72,** 35.

4. SONNEBORN, T. M. (1970). *Proc. Roy. Soc. Lond.* B. **176,** 347.

5. WRIGHT, B. E. (1966). *Science* **153,** 830.

1

The Importance of Metabolite Flux

> "... *the fact that intermediate products do not usually accumulate*
> *shows that the substrates of the intermediary enzymes are removed*
> *as rapidly as they are formed. The average half-life of the acids*
> *of the tricarboxylic acid cycle in a rapidly respiring tissue is*
> *a few seconds. The amounts of enzyme in the tissue are sufficient*
> *to deal with the intermediates as soon as they arise; in other words,*
> *the amount of available substrate is the factor limiting the rate*
> *at which the intermediary step proceeds.*"
>
> H. A. Krebs, 1957 [51]

Metabolite Flux Control of Reaction Rates

Atkinson has recently pointed out that the existence in cells of thousands of different polyelectrolytes and smaller molecules poses serious problems of solubility [4]. It may, therefore, have been necessary in the evolutionary development of metabolism to select for effective mechanisms limiting the concentration of metabolites. Krebs [51, 52] and many others [9, 56, 89, 93] have often emphasized that, although there are a few "pacemaker" enzymes such as phosphorylase and hexokinase, substrates are generally more limiting than enzymes in the intact cell. Changes in the rate of their availability should, therefore, exert significant effects on the rates of reactions essential to differentiation. In ascites tumor cells, glycolysis is limited primarily by the concentration of intracellular phosphate [105], and respiration appears to be limited by the level of adenosine diphosphate (ADP) [16]. Control *in vivo* of metabolic activity by the flux of metabolites into a pathway

8

has been intensively studied, especially in the case of glycolysis. The conclusion has been reached that enzyme profiles have little functional significance and are a poor indication of the metabolic state. The important parameter is the net flux through a metabolic pathway. This parameter, along with the steady state levels of the metabolites, gives the steady state flux pattern, reveals the relationships or "kinetic positions" of the enzyme reactions, and thus indicates the control points in the pathways concerned [37].

An illustration of the relative importance of changes in metabolite flux as compared to enzyme levels may be seen in the studies of Kim and Miller with intact and adrenalectomized rats and with perfused liver [47]. The rate of $^{14}CO_2$ formation from tryptophan-3-[^{14}C] or tyrosine-1-[^{14}C] was followed before and after hydrocortisone administration, which results in manifold increases in the activities of tryptophan oxygenase and tyrosine transaminase in both intact and perfused livers. These increased enzyme levels did not result in parallel increases in the oxidation of [^{14}C-] labeled amino acids to $^{14}CO_2$. However, increasing the substrate load *per se* was associated with large increases in the rate of amino acid oxidation, under conditions not associated with altered enzyme levels. In another system, the induction of nitrate reductase by nitrate in intact barley tissue, studies indicate that only a small fraction of the induced enzyme is functioning *in vivo* [23].

In some cases, depending upon the position of an enzyme in a sequence of coupled reactions, changes in its activity may have profound effects on the rate of substrate utilization. The work of Kascer provides an interesting example of the sensitivity of metabolite flux in steady state systems to changes in the level of an enzyme, depending on its relative ("kinetic") position in a metabolic sequence [44]. Using the pathway of arginine synthesis in *Neurospora* as an example, Kascer has analyzed the relationship between the enzyme controlling the final step in arginine formation, argininosuccinase (ASAase), and the extent of arginine accumulation:

$$ORN \xrightarrow{\text{OTCase}} CIT \xrightarrow{\text{SYNase}} ASA \xrightarrow{\text{ASAase}} ARG \longrightarrow Protein$$

In a simplified mathematical treatment of this system, enzyme concentrations were expressed as rate constants and a set of simultaneous first-order differential equations were formulated. The specific activity of ASAase could be varied by the use of mutants. An analysis of the coupled reactions involved in such a system predicted that the steady-state value of arginine would be relatively independent of the enzyme ASAase. In confirmation of this prediction, strains with striking differences in this enzyme activity showed no significant differences in arginine pool sizes. On the other hand, Kascer's analysis indicated that changes in the catalytic activity as this step should be accompanied by appropriate changes in the steady-state precursor concentrations. Their enhanced accumulation should compensate for a lowered catalysis and vice versa. As predicted, an inverse correlation was observed between specific enzyme activity and the concentration of argininosuccinate (ASA, see Fig. 2). While some enzymes (e.g., ASAase) had no effect on the

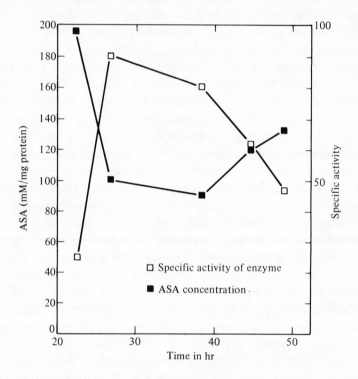

Figure 2. Determination of argininosuccinate pools and specific activity of ASAase in mycclium of the mutant $r - 1$. [44]

metabolite flux through the pathway, the theoretical analysis predicted that the first enzyme in the series, ornithine transcarbamylase (OTCase) should be flux controlling, because of its kinetic position—i.e., because it controlled the flow of material into the metabolic sequence. Again, a mutant having 2% of the normal activity for this enzyme had 10% of the normal arginine value. Another interesting case in *Neurospora* has been described by Brody and Tatum [11]. A mutation resulting in a decrease in the affinity of glucose-6-phosphate (G6P) dehydrogenase for both G6P and nicotinamide adenine dinucleotide phosphate (NADP) results in a tenfold higher steady state level of G6P compared to the wild type. Kascer emphasizes that a set of enzymes, coupled by their shared substrates, acts as a unit in determining the flux. Within this unit, changes in catalytic activity have negligible effects; control is mainly exercised by the enzymes initiating (e.g., OTCase) and terminating (e.g., an enzyme involved in the subsequent metabolism of arginine) the set. This conclusion is consistent with the observation that feedback control also acts on enzymes initiating a series of reactions [5].

Since many primitive systems differentiate under starvation conditions and must depend upon limited amounts of material to use for chemical energy and precursors in synthetic reactions (see Chap. 4), changes in metabolite flux might be expected to play an even more important role than in cells not limited by nutrient availability. Unfortunately, very little information is available in differentiating systems on substrate levels or flux values determined *in vivo*. Work with *Arbacia* eggs from Krahl's laboratory in 1955 was concerned with the factors which limit oxygen consumption in the unfertilized and fertilized egg [50]. During development, eggs depend in large part upon carbohydrate oxidation as a source of energy with G6P oxidation by NADP as the major pathway. Using G6P as substrate and a supernatant fraction from eggs as a source of enzymes, these investigators found that the rate of NADP reduction was sufficient to support a rate of oxygen consumption 24 times that observed for unfertilized and six times that for fertilized eggs. These data suggested a limiting level of hexose phosphates; indeed, some ten years later Aketa, et al., determined the levels of these esters in *Arbacia* eggs and found them to be limiting for respiration [1]. They considered it probable that a restriction existed in the pathway leading from glycogen to G6P; phosphorylase activation was suggested as the likely rate-limiting reaction.

In the cellular slime mold, *D. discoideum*, differentiation occurs under starvation conditions, and endogenous materials are limiting for both synthetic and catabolic pathways. Many specific reactions which have been examined appear to be substrate-limited, judging from the relative concentration of substrates in the cell and the affinity of their respective enzymes for them—i.e., the substrate/Km ratio is low. Experiments *in vivo* carried out with radioactive tracers have demonstrated that glutamate oxidation [96], glucose catabolism [102], UDPG synthesis [69, 70], and glycogen synthesis [61, 74, 103] are among the reactions limited by endogenous substrate availability. In each case, as we shall see, these substrates accumulate during differentiation. Therefore, their accumulation and increased rate of flux would enhance the rate of the reaction. Thus, flux control may be of primary importance in *D. discoideum;* the next two chapters will treat this model system in detail.

Metabolite Flux Control of Enzyme Levels

Introduction

The observation that enzyme activities change over the course of differentiation and the probable importance of these changes have been the subject of extensive investigations. Recent reviews are available dealing with this aspect of differentiation in amphibia [95], mammals [34], insects [31], plants [24, 68], and microbes [14, 67, 76, 78, 98]. A central dogma of developmental biology today appears to be that differentiation will follow only if the right enzyme is made at the right place and right time. A basic assumption underlying this dogma is that alterations in the concentration of specific enzymes will result in, that is, be critical to an altered rate of synthesis of some material essential to differentiation. This is an assumption we wish to challenge in subsequent chapters. By analogy to bacterial systems, changing enzyme patterns during differentiation are usually thought to involve an altered rate of enzyme synthesis and to be initiated at the nucleic acid level. This is undoubtedly true in cases involving a response of the system to hormones since the latter are known to affect enzyme levels in mammals by altering their rate of synthesis. Thus, studies in rat liver have shown that tyrosine transaminase [45], glutamate-alanine transaminase [86], tryptophan oxygenase [83], and serine dehydratase [42] increase in activity following hormone administration. That an enhanced rate of enzyme synthesis is the responsible mechanism was demonstrated by a combination of isotope incorporation and immunochemical studies. (For an excellent review, see Schimke and Doyle [82].) Presumably hormone-induced increases of enzyme activities in fetal rats also involve alterations in the rate of enzyme synthesis [34]. The effective hormones, glucagon and epinephrine, are probably the natural stimuli for the "spontaneous" enzyme accumulation occurring normally after birth. The mechanism(s) by which hormones exert an effect on the rate of enzyme synthesis is not yet clear; many possibilities exist among the events upon which protein synthesis depends. These include the synthesis of mRNA, the conversion of preexisting RNA to mRNA by stabilization or transport [36], the translation of mRNA [27], or the regulation of ribosome function [43]. The latter mechanism has recently been implicated in the activation of protein synthesis following fertilization of sea urchin eggs [63]. Another mechanism, known to occur in differentiating systems, is the activation of preexisting (masked) mRNA [17, 64, 88]. Although not relevant to changes in the rate of enzyme synthesis, the activation of masked enzymes in developing plants and animals are probably of great importance as control mechanisms during differentiation [6, 20–22, 35, 62, 92].

In contrast to the enzymes in rat liver, cited above, an enhanced rate of enzyme synthesis has not been unequivocally demonstrated in a differentiating system. In a number of cases—e.g., α-amylase accumulation in response to gibberellic acid [24], an enzyme has been shown to arise from amino acids and not a preformed precursor. Although an increased rate of synthesis is probably the responsible mechanism, an unaltered rate of enzyme synthesis coupled with decreased degradation was not excluded (see next section).

In mammalian systems a number of investigations [8, 26, 33, 58, 80] have suggested that, in contrast to hormones, substrates and cofactors appear to affects enzyme levels by altering their rate of degradation. This latter mechanism is therefore of particular interest in a monograph concerned with metabolite effects and could prove to be of general importance in differentiating systems, especially in view of the fact that the concentration of substrates, cofactors, and modifiers may undergo striking changes over the course of differentiation [7, 18, 65, 99]. If differentiating organisms (including microbial systems) are metabolically more analogous to higher organisms than to exponentially-growing bacteria, the role of metabolites in affecting the rate of enzyme degradation may be as relevant as their role in enzyme induction (i.e., increased rate of enzyme synthesis). Prior to a discussion of cases in differentiating systems which may illustrate the effects of metabolites on enzyme degradation, it is useful to briefly review selected studies in systems not undergoing active differentiation.

Control of enzyme levels through changes in the rate of degradation in differentiated systems

In exponentially-growing bacteria little protein degradation can be demonstrated [41]. In contrast, protein breakdown and resynthesis (turnover) occurs in systems which are growing at a low rate or not at all [80, 85, 91, 101]. The extensive protein turnover observed in nongrowing systems is presumably required for the maintenance of enzymes essential for the general metabolic requirements of the cell. In systems dependent upon endogenous metabolism (including primitive differentiating systems), protein degradation would also be a necessary source of precursor material for the synthesis of any specialized enzymes that might be formed:

$$\text{Proteins} \xrightarrow{\text{deg.}} \begin{array}{c}\text{Peptides} +\\ \text{amino acids}\end{array} \xrightarrow[\text{A}]{\text{syn.}} \text{Enzyme} \xrightarrow[\text{B}]{\text{deg.}} \begin{array}{c}\text{Peptides} +\\ \text{amino acids}\end{array}$$

As shown here, the accumulation of a particular enzyme with time could occur either by a stimulation of its synthesis (point A) or by inhibition of its breakdown (point B). The latter can be achieved by stabilization of the enzyme by its substrate, a widely-documented phenomenon [2, 8, 32, 33, 84]. Since the steady-state level of an enzyme is the result of a balance between its rates of synthesis and degradation, *alterations* in enzyme level may be due to changes in the rate of either degradation or synthesis.

Evidence is now available indicating that enzyme accumulation *in vivo* may result from retarding enzyme degradation. The enhanced levels of tryptophan oxygenase in response to tryptophan, for example, is achieved by substrate stabilization of the enzyme against degradation *in vivo*, not by an increase in the rate of enzyme synthesis [83]. This study as well as a number of others [8, 26, 33, 58] has led Schimke to conclude that degradation is equally as important as protein synthesis in regulating the amount of an enzyme [80]. Starvation [81] or changes in diet [53, 58] are also known to affect the rate of enzyme degradation, perhaps through alterations in levels of endogenous metabolites. Studies with these systems may help elucidate the mechanisms underlying changes in metabolite and enzyme levels known to occur during differentiation, which frequently is also accompanied by starvation (see Chap. 4).

Another example of the epigenetic control of enzyme levels may be seen in studies with isozymes. In contrast to previous interpretations, it has been shown recently that tissue-specific patterns of lactate dehydrogenase (LDH) isozymes do not arise solely from differential activation of the genes controlling formation of the subunits. Wide differences among tissues in the rate of isozyme degradation suggest that concentrations of various metabolites shown to affect isozyme decay *in vitro* might play a role in regulating their catabolism *in vivo* [10, 26]. The synthetic and degradative rate constants were compared for the LDH-5 isozymes of three rat tissues. The rate of enzyme synthesis varied as much as 30-fold, the rate of degradation as much as 20-fold; the relationship between synthesis and degradation also changes from one tissue to another [26].

When a cell can no longer grow (as often happens during differentiation), enzymes may become nonfunctional to varying degrees due to allosteric modification or lack of substrate. They then may become vulnerable to catabolism. Consistent with this conclusion is the fact that a number of correlations exist between the stability of an enzyme *in vitro* and its stability *in vivo* [8, 33, 84]. The model thus emerges that, depending upon shifting distributions and concentrations of substrates, cofactors, etc., specific enzymes are stabilized or labilized to attack by degradative enzymes which may be present at all times in excess (for a discussion, see Schimke [80] and Pine [71]).

Inhibitor studies

An inhibition of enzyme accumulation by actinomycin D has frequently been used as evidence that changes only in the rate of enzyme synthesis are involved. However, Reel and Kenney have demonstrated that both synthesis and degradation can be inhibited by actinomycin D [73] or cycloheximide [46]. This makes it extremely difficult to arrive at conclusions regarding template stability from inhibitor studies. If synthesis and degradation are inhibited to about the same extent, the level of an enzyme may remain unchanged, yet not indicate the presence of a stable mRNA. In one case, tyrosine transaminase, actinomycin D inhibits degradation more than synthesis, resulting in an actual increase in enzyme level ("super-induction") [73].

As noted earlier in the case of tryptophan oxygenase, enzyme accumulation as a function of time can be due only to a decreased rate of degradation resulting from the addition of stabilizing substrate. If the accumulation of endogenous substrates and products during differentiation results in an analogous effect, enzyme accumulation could be due to a decreased rate of degradation alone. Regardless of whether this mechanism or changes in the rate of enzyme synthesis are responsible for enzyme accumulation, actinomycin D will act as an inhibitor.

Control of enzyme levels through changes in rate of degradation in differentiating systems

Turning now to differentiating systems, a number of studies in higher plants suggest that substrate stabilization may be the underlying mechanism responsible for alterations in enzyme levels as a function of time. Of particular interest are cases in which the addition of exogenous metabolites further enhances enzyme activity. This occurs in lily microspores for thymidine kinase [39], an enzyme known to be stabilized by its substrate [8], and also occurs with the enhanced activity of hexokinase and fructokinase in castor bean cotyledons [60]. In both systems actinomycin D inhibits the increase

of enzyme activity, but as we have just seen, this does not necessarily implicate alterations in the rate of enzyme synthesis as the responsible mechanism.
For example, the drug could interfere with enzyme accumulation by preventing the accumulation of stabilizing endogenous metabolites or inhibiting the
synthesis of a proteolytic enzyme responsible for unmasking the enzyme of
interest, or by a number of other mechanisms. In germinating peanuts
isocitratase and malate synthetase have been shown to arise "de novo" by
the density-labeling method, which again does not distinguish between
changes in the rate of synthesis or degradation or both as the mechanism for
enzyme accumulation [24]. In many of these cases, removal of the metabolite
responsible for altered enzyme levels results in a decline of enzyme activity, an observation which is compatible with a substrate stabilization mechanism [80]. Such a mechanism may also account for enhanced cholinesterase
levels following the addition of acetylcholine to cultures of chick embryonic
cells [13], for increased alcohol dehydrogenase activity following the addition
of ethanol to embryonic cells of *Drosophila* [38], or for increased levels of
alkaline phosphatase in primary fibroblasts from human skin when β-gly-
cerophosphate is added to the medium [19]. Changes in the level of UDP
pyrophosphorylase in *D. discoideum* may involve a decrease in the rate of
enzyme degradation during differentiation; this possibility is discussed in
Chap. 2.

Control of enzyme levels through changes
in the rate of synthesis (induction)

Changing levels of endogenous metabolites could induce the synthesis of
enzymes during differentiation by a mechanism analogous to the induction
of enzymes in bacteria. However, direct evidence, such as that available for
β-galactosidase in *E. coli* has not yet been obtained for metabolite-induced
enzyme synthesis during differentiation. In considering the role of metabolites in the induction and repression of enzyme synthesis, perhaps the most
appropriate system to mention is bacterial sporulation. Since a number of
recent reviews on this subject are available [14, 78], only a few examples shall
be cited here.

Unlike more complex examples of differentiation in higher organisms,
bacterial sporulation can be modulated by changing environmental conditions and is not exclusively dependent upon metabolic changes within the
system (e.g., an avian egg). In *Bacillus* protease activity increases rapidly
at the end of growth by "de novo" synthesis. Both sporulation and protease

activity can be repressed by the addition of high levels of amino acids [15, 54, 66], and one of the earliest changes during starvation in some microbes is a reduction in level of endogenous amino acids [25, 59, 100]. If the appearance of protease activity were normally repressed by endogenous amino acids, the synthesis of such enzymes could be initiated on starvation, which is frequently a prerequisite for sporulation. A comparable situation exists for three enzymes involved in arginine degradation (arginase, ornithine-δ-transaminase, and Δ'-pyrroline-5-carboxylate dehydrogenase). These enzymes normally appear only during sporulation, at a time when glucose has been exhausted from the medium. If cells are grown in the absence of glucose, these enzymes are present prior to sporulation, suggesting catabolite repression by glucose [53, 79]. Citrate synthase and aconitase appear to respond to both catabolite and feedback repression. In *Bacillus licheniformis* a positive correlation exists between the level of cellular metabolites and the activity of threonine dehydrase, aspartokinase, and pyruvate kinase [7]. Conditions preventing a decrease in endogenous metabolite level at the end of growth also prevented a loss in enzyme activity, a situation reminiscent of earlier discussions. In this case, however, the authors suggest an allosteric modification of the enzyme by intracellular metabolites as a possible mechanism for affecting enzyme activity.

In the cellular slime mold, *Dictyostelium discoideum*, an example is known in which endogenous metabolism results in the accumulation of a metabolite which represses the formation of an enzyme during differentiation. The intracellular concentration of inorganic phosphate (Pi) increases tenfold as the cells starve, reaching concentrations of 0.05 M [30]. Since in this organism a 5'-nucleotidase as well as an acid phosphatase [29] can be repressed by the addition of comparable levels of exogenous Pi, enzyme repression may normally occur *in vivo* by the accumulation of endogenous Pi. Another possible example, in developing chick embryos [94], is the repression by creatine of arginine-glycine transamidase. Thus, a general mechanism for the accumulation of enzymes unique to differentiation may be the appearance or disappearance on starvation of specific inducers or corepressors affecting the synthesis of these enzymes. The role of "chemical-messengers" in the activation and repression of gene expression during differentiation has been treated in an interesting manner by Bullough [12].

We have seen that alterations in metabolite flux could affect the concentration of enzymes at the site of their synthesis or degradation. Regardless of the mechanism, relatively little attention has been paid to the source of these metabolites in differentiating systems. In investigations of enzyme induction in bacterial systems the inducer is added to the cell culture. By contrast, primitive differentiating systems are usually isolated from external metabolites, as will be seen in Chap. 4. Changes in endogenous metabolism must, therefore, be responsible for changes in the population of metabolites essential for enzyme induction or stabilization. In view of this, the ultimate source of these molecules and the control of their synthesis and metabolism are as essential to enzyme accumulation as the presence of the enzyme or gene itself.

There are of course other mechanisms by which enzyme levels change that do not appear to involve direct interaction with metabolites. Examples include hormone activation, discussed earlier, the activation of masked mRNA [64, 88], or the transfer of an enzyme from one location in the cell to another with consequent changes in metabolic activity [40, 104]. The release of active enzyme from an inactive precursor during differentiation appears to be a common mechanism [6, 20–22, 35, 62, 92]; cases in the cellular slime mold will be discussed below.

Metabolite Flux Control of Enzyme Activity
by Allosteric Modulation and End-product Inhibition

The amplification effects on the activity of an enzyme which can occur at the level of positive and negative modifiers and multiple substrates suggest that this type of control will eventually be recognized as very important in affecting the activity of metabolic pathways underlying differentiation. A useful example to illustrate the potential effect of multiple modifiers on the activity of an enzyme can be seen in the case of phosphofructokinase [28, 87]. The large number of positive and negative modifier molecules involved allows relatively sharp activation and cutoff points. At such points enzyme activity can change significantly in response to a very moderate alteration in the concentration of the appropriate modifiers. Thus, the sudden appearance of a new product during differentiation may represent only the culminating effect of many minor, interacting changes in metabolism. Extensive discussions of these aspects of metabolic control may be found in a number of excellent reviews [3–5, 77, 90]; some of these discussions also deal with the more subtle changes in an enzyme's environment which could affect its conformation or active concentration [48, 49]. A very interesting example of the modulation of enzyme activity by metabolites may be seen in the effects of citrate on the activity, conformation, and size of acetyl CoA carboxylase [57].

Few examples of this type of metabolite control are as yet available for differentiating systems, as changes in the level of endogenous metabolites affecting relevant pathways are rarely studied. A few cases are known in *D. discoideum*. As will be seen, changes in the level of the modifier, G6P, partly control the relative utilization of limiting levels of UDPG with respect to the synthesis of two types of polysaccharide [99]. Also, 3'5' AMP and 5' AMP both stimulate glycogen degradation by different mechanisms (see Chap. 2) Inorganic phosphate accumulates late in differentiation to levels which would significantly inhibit the activity of a specific 5'-nucleotidase and an acid phosphatase *in vivo* [29, 30], as well as the synthesis of soluble glycogen, cell wall glycogen and trehalose [75]. Some of these effects are indicated in Table 1. The levels of Pi which accumulate in the terminal stages of differentiation may thus be instrumental in shutting down synthetic activities and sparing endogenous materials for the maintenance and subsequent germination of the spores. As will be discussed in the following chapter, two other metabolites accumulating to inhibitory levels are UDPG, which affects both UDPG pyrophosphorylase and glycogen phosphorylase, and G1P, which inhibits glycogen phosphorylase.

Table 1

THE EFFECT OF PI ON FOUR ENZYMES OF D. DISCOIDEUM
PERCENT INHIBITION

Pi (mM)[a]	5'nucleotidase	Acid P'ase[b]	Soluble glycogen synthetase	Cell wall glycogen synthetase
0.2			39.0	
2.0		50.0	70.0	7.0
3.0	25.0	70.0	70.0	12.0
6.0	45.0	85.0	72.0	17.0
20.0	78.0	100.0	73.0	28.0
60.0	90.0	100.0	80.0	50.0

[a] *The range in vivo is from 3.0 to 50.0 expressed in terms of packed cell volume* [30].
[b] *Using G6P as substrate* [29].

Metabolite Flux Control of the Initiation, Duration, Termination, and Stability of Differentiation

In this context we are largely concerned with the ultimate source of metabolites: the macromolecular reserves on which many systems depend for survival and differentiation. This subject will be treated in more detail in the last chapter. Each system starts with a reproducible and fixed amount of

specific reserve material, located within each cell or exterior to the organism
—e.g., the yolk of eggs. These endogenous reserves are for the most part
used in an orderly and sequential manner. In effect, a "programmed starvation" occurs, at the end of which differentiation has been accomplished
[97]. The appearance of hydrolytic enzymes such as proteases, phosphopro-
tein phosphatases, and lipases is therefore of fundamental importance to an
understanding of substrate control in differentiation.

There is a limited amount of reserve material to be used as a source of
(1) energy and (2) precursors for the maintenance of essential macromolecules
(turnover) as well as for the synthesis of new enzymes and products which
accumulate during morphogenesis. This amount is fixed in concentration by
the composition of the organism in question and in part defines the period
of time over which differentiation can occur. Dependence upon a specific
amount of reserve material must also be a determining factor in controlling
the pattern of appearance of the intermediates and end products accumulating at particular stages of differentiation. At a more detailed, mechanistic
level, it has been found that many of the changes in metabolite levels associated with and probably resulting from the depletion of specific endogenous
reserves may actually help to initiate and terminate differentiation. For
example, during starvation and differentiation in *D. discoideum*, oxidizable
reserves are partially depleted and oxygen consumption decreases. A lowered
rate of oxidative phosphorylation may well result in Pi accumulation, which
was prevented earlier due to the rapid esterification of Pi to ATP. As discussed above, cellular levels of Pi may help establish and maintain the
dormant state of the spores.

A later discussion will expand on the observation that the flux of precursor
material through a biosynthetic pathway will be largely controlled by the
enzyme catalyzing the reaction resulting in the appearance of the first substrate in the sequence. Reproducible changes in the levels of endogenous
metabolites offer a mechanism by which the activity of specific synthetic
pathways could be elicited and maintained: Availability of the first substrate
in a series of reactions would enhance the metabolite flux of intermediates,
with a consequent increase in enzyme levels by various mechanisms, insuring
the stability of the pathway. Referring to glycolysis Racker speculates that
an excess of enzymes at important crossroads of metabolism may provide
a "metabolic buffer" not readily disturbed by small changes in catalytic
activities [72]. An increased enzyme level may or may not result in an
enhanced reaction rate *in vivo*. As will be seen in Chap. 3, an increase in enzyme level may also serve to compensate for the accumulation of an end product inhibitor or to anticipate the need for a rapid and efficient utilization
of substrate at some future time in the course of differentiation.

The patterns of enzyme-substrate-product-modifier-interaction which have evolved in the starving slime mold system confer enormous flexibility since, in general, a variation in the concentration of any one metabolite merely changes the operating range of several others rather than rigidly controlling the end result [99]. The intracellular concentrations of essential metabolites vary significantly from one group of cells to another—yet normal differentiation always occurs; this implies great versatility in the mechanisms underlying the activity and interaction of the various metabolic pathways involved. Such a flexibility may be fundamental to the stability and reproducibility of the transformations which result in differentiation.

Very few differentiating systems have been analyzed in sufficient depth to permit a consideration of the role of metabolite flux in the control of differentiation. One such system is the cellular slime mold, which will now be described in detail.

REFERENCES

1. AKETA, K., BIANCHETTI, R., MARRÈ, E., and MONROY, A. (1964). *Biochem. et Biophys. Acta* **86,** 211.

2. ALPERS, J. B., WU, R., and RACKER, E. (1963). *J. Biol. Chem.* **238,** 2274.

3. ATKINSON, D. E. (1969). *Ann. Rev. Microbiology* **23,** 47.

4. ATKINSON, D. E. (1969). In *Current Topics in Cellular Regulation* I (B. L. Horecker and E. R. Stadtman, eds.). New York: Academic Press Inc., p. 29.

5. ATKINSON, D. E. (1965). *Science* **150,** 851.

6. BERG, W. E. (1950). *Biol. Bull.* **98,** 128.

7. BERNLOHR, R. W. and GRAY, B. H. (1969). In *Spores IV* (L. L. Campbell, ed.). Bethesda, Md.: American Society for Microbiology, p. 186.

8. BOJARSKI, T. B. and HIATT, H. H. (1960). *Nature* **188,** 1112.

9. BOWMAN, R. H. (1966). *J. Biol. Chem.* **241,** 3041.

10. BOYD, J. W. (1967). *Biochim. et Biophys. Acta* **132,** 221.

11. BRODY, S. and TATUM, E. L. (1966). *Proc. Nat. Acad. Sci. U. S.* **56,** 1290.

12. BULLOUGH, W. S. (1967). *The Evolution of Differentiation.* New York: Academic Press Inc.

13. BURKHALTER, A., JONES, M., and FEATHERSTONE, R. M. (1957). *Proc. Soc. Exptl. Biol. Med.* **96,** 747.

14. CAMPBELL, L. L. (ed., 1969). *Spores IV.* Bethesda, Md.: American Society for Microbiology.

15. CHALOUPKA, J., KRECKOVA, P., and RIBOVA, L. (1963). *Biochem. Biophys. Res. Commun.* **12,** 380.

16. CHANCE, B. and HESS, B. (1959). *Science* **129**, 700.

17. CHEN, D., SAVID, S., and KATCHALSKI, E. (1968). *Proc. Nat. Acad. Sci. U. S.* **60**, 902.

18. CLELAND, S. V. (1969). Ph.D. Dissertation "Gluconeogenesis and Glycolysis in *Dictyostelium discoideum*," Northwestern University.

19. COX, R. P. and PONTECORVO, G. (1961). *Proc. Nat. Acad. Sci. U. S.* **47**, 839.

20. DAGGS, R. G. and HALCRO-WARDLAW, H. S. (1933). *J. Gen. Physiol.* **17**, 303.

21. EPEL, D., WEAVER, A. M., MUCHMORE, A. V., and SCHIMKE, R. T. (1969). *Science* **163**, 294.

22. ETZLER, M. E. and MOOG, F. (1966). *Science* **154**, 1037.

23. FARRARI, T. E. and VARNER, J. E. (1970). *Proc. Nat. Acad. Sci. U. S.* **65**, 729.

24. FILNER, P., WRAY, J. L., and VARNER, J. E. (1969). *Science* **165**, 358.

25. FOSTER, J. W. and PERRY, J. J. (1954). *J. Bact.* **67**, 295.

26. FRITZ, P. J., VESELL, E. S., WHITE, E. L., and PRUITT, K. M. (1969). *Proc. Nat. Acad. Sci. U. S.* **62**, 558.

27. FUHR, J. E., LONDON, I. M., and GRAYZEL, A. I. (1969). *Proc. Nat. Acad. Sci. U. S.* **63**, 129.

28. GARFINKEL, D. (1965). *J. Biol. Chem.* **241**, 286.

29. GEZELIUS, K. (1966). *Physiol. Plant.* **19**, 946.

30. GEZELIUS, K. and WRIGHT, B. E. (1965). *J. Gen. Microbiol.* **38**, 309.

31. GILBERT L. I. (1967). In *Comprehensive Biochemistry*, **28**: *Morphogenesis, Differentiation and Development* (M. Florkin and E. H. Stotz, eds.). New York: Elsevier Publishing Co.

32. GREEN, N. M. and NEURATH, H. (1954). In *The Proteins* **2** (H. Neurath and K. Bailey, eds.). New York: Academic Press Inc.

33. GREENE, M. L., BOYLE, J. A., and SEEGMILLER, J. E. (1970). *Science* **167**, 887.

34. GREENGARD, O. (1969). *Science* **163**, 891.

35. HARRIS, D. L. (1946). *J. Biol. Chem.* **165**, 541.

36. HARRIS, H. and WATTS, J. W. (1962). *Proc. Roy. Soc. London, B.* **156**, 109.

37. HESS, B. and BRAND, K. (1965). In *Control of Energy Metabolism* (B. Chance, R. W. Estabrook, and J. R. Williamson, eds.). New York: Academic Press Inc.

38. HORIKAWA, M., LING, L. L., and FOX, A. S. (1967). *Genetics* **55**, 569.

39. HOTTA, Y. and STERN, H. (1965). *J. Cell. Biol.* **25**, 99.

40. ISONO, N. (1963). *J. Fac, Sci. Univ. of Tokyo* **10**, 37.

41. JACOB, F. and MONOD, J. (1961). *J. Mol. Biol.* **3**, 318.

42. JOST, J. O., KHAIRALLAH, E. A., and PITOT, H. C. (1968). *J. Biol. Chem.* **243**, 3057.

43. KANO-SUEOKA, T. and SUEOKA, N. J. (1966). *J. Mol. Biol.* **20,** 183.

44. KASCER, H. (1963). In *Biological Organization at the Cellular and Super-cellular Level* (R. J. C. Harris, ed.). New York: Academic Press Inc.

45. KENNEY, F. T. (1962). *J. Biol. Chem.* **237,** 3495.

46. KENNEY, F. T. (1967). *Science* **156,** 525.

47. KIM, J. H. and MILLER, L. L. (1968). *J. Biol. Chem.* **244,** 1410.

48. KOSHLAND, D. E., JR. and KIRTLEY, M. E. (1966). In *Major Problems in Developmental Biology, 25th Symposium of the Society for Developmental Biology* (M. Locke, ed.). New York: Academic Press Inc.

49. KOSHLAND, D. E., Jr. (1969). In *Current Topics in Cellular Regulation* I (B. L. Horecker and E. R. Stadtman, eds.). New York: Academic Press Inc., p. 1.

50. KRAHL, M. E., KELTCH, A. K., WALTERS, C. P., and CLOWES, G. H. A. (1955). *J. Gen. Physiol.* **38,** 431.

51. KREBS, H. A. (1957). *Endeavour XVI*, 125.

52. KREBS, H. A. (1969). In *Current Topics in Cellular Regulation* I (B. L. Horecker and E. R. Stadtman, eds.). New York: Academic Press Inc., p. 45.

53. LAISHLEY, E. J. and BERNLOHR, R. W. (1968). *J. Bact.* **96,** 322.

54. LEVISOHN, S. and ARONSON, A. I. (1967). *J. Bact.* **93,** 1023.

55. LOFTFIELD, R. B. and HARRIS, A. (1956). *J. Biol. Chem.* **219,** 151.

56. LOWRY, O. and PASSONNEAU, J. V. (1964). *J. Biol. Chem.* **239,** 31.

57. LYNEN, F. (1970). In *Control Processes in Multicellular Organisms* (G. E. W. Wolstenholme and Julie Knight, eds.). Ciba Foundation Symposium. Boston, Mass.: Little, Brown and Company.

58. MAJERUS, P. W. and KILBURN, E. (1969). *J. Biol. Chem.* **244,** 6254.

59. MANDELSTAM, J. (1958). *Biochem. J.* **69,** 110.

60. MARRÈ, E., CORNAGGIA, M. P., ALBERGHINA, F., and BIANCHETTI, R. (1965). *Biochem. J.* **97,** 20P.

61. MARSHALL, R., SARGENT, D., and WRIGHT, B. E. (1970). *Biochemistry* **9,** 3087.

62. MAYER, A. M. and SHAIN, Y. (1968). *Science* **162,** 1283.

63. METAFORA, S., FELICETTI, L., and GAMBINO, R. (1971). *Proc. Nat. Acad. Sci.* **68,** 600.

64. MONROY, A., MAGGIO, R., and RINALDI, A. M. (1965). *Proc. Nat. Acad. Sci. U. S.* **54,** 107.

65. NELSON, D. L., SPUDICH, J. A., BONSEN, P. P. M., BERTSCH, L. L., and KORNBERG, A. In *Spores IV* (L. L. Campbell, ed.). Bethesda, Md.: American Society Microbiology, p. 59.

66. NEUMARK, R. and CITRI, N. (1962). *Biochim. et Biophys. Acta* **59,** 749.

67. NIEDERPRUEM, D. J. and WESSELS, J. G. H. (1969). *Bacteriol. Rev.* **33,** 505.

68. OVERBEEK, J. VAN (1966). *Science* **152,** 721.

69. PANNBACKER, R. G. (1967). *Biochemistry* **6,** 1283.

70. PANNBACKER, R. G. (1967). *Biochemistry* **6,** 1287.

71. PINE, M. J. (1966). *J. Bact.* **92,** 847.

72. RACKER, E. (1965). In *Mechanisms in Bioenergetics.* New York: Academic Press Inc., p. 213.

73. REEL, J. R. and KENNEY, F. T. (1968). *Proc. Nat. Acad. Sci. U. S.* **61,** 200.

74. ROSNESS, P. A., GUSTAFSON, G., and WRIGHT, B. E. (1971). *J. Bact.* **108,** 1329.

75. ROTH, R. and SUSSMAN, M. (1968). *J. Biol. Chem.* **243,** 5081.

76. RUSCH, H. P. (1969). *Fed. Proc.* **28,** 1761.

77. SANWAL, B. D. (1970). *Bact. Rev.* **34,** 20.

78. SCHAEFFER, P. (1969). *Bact. Rev.* **33,** 48.

79. SCHAEFFER, P., MILLET, J., and AUBERT, J. (1965). *Proc. Nat. Acad. Sci. U. S.* **54,** 704.

80. SCHIMKE, R. T. (1969). In *Current Topics in Cellular Regulation* I (B. L. Horecker and E. R. Stadtman, eds.). New York: Academic Press Inc., p. 77.

81. SCHIMKE, R. T. (1964). *J. Biol. Chem.* **239,** 3808.

82. SCHIMKE, R. T. and DOYLE, D. (1970). "Control of enzyme levels in animal tissues." In *Ann. Rev. Biochem.* **39,** 929.

83. SCHIMKE, R. T., SWEENEY, E. W., and BERLIN, C. M. (1965). *J. Biol. Chem.* **240,** 322.

84. SCHIMKE, R. T., SWEENEY, E. W., and BERLIN, C. M. (1965). *J. Biol. Chem.* **240,** 4609.

85. SCHOENHEIMER, R. (1942). *The Dynamic State of Body Constituents.* Cambridge, Mass.: Harvard University Press.

86. SEGAL, H. L. and KIM, Y. S. (1963). *Proc. Nat. Acad. Sci. U. S.* **50,** 912.

87. SHEN, L. C., FALL, L., WALTON, G. M., and ATKINSON, D. E. (1968). *Biochemistry* **7,** 4041.

88. SPIRIN, A. S. (1966). In *Current Topics in Developmental Biology* I New York: Academic Press Inc.

89. SRERE, P. A. (1967). *Science* **158,** 936.

90. STADTMAN, E. R. (1963). *Bact. Rev.* **27,** 170.

91. SWICK, R. W. (1958). *J. Biol. Chem.* **231,** 751.

92. TATIBANA, M. and COHEN, P. P. (1965). *Proc. Nat. Acad. Sci. U. S.* **53,** 104.

93. VETGOSKY, A. and FRIEDEN, E. (1958). *Enzymologia* **19,** 143.

94. WALKER, J. G. (1963). In *Advances in Enzyme Regulation.* London: Pergamon Press Ltd.

95. WEBER, R. (1967). In *Comprehensive Biochemistry*, **28**: *Morphogenesis, Differentiation, and Development* (M. Florkin and E. H. Stotz, eds.). New York: Elsevier Publishing Co.

96. WRIGHT, B. E. (1963). *Bact. Rev.* **27**, 273.

97. WRIGHT, B. E. (1964). In *Comparative Biochemistry* **6** (M. Florkin and H. S., Mason, ed.). New York: Academic Press Inc.

98. WRIGHT, B. E. (1968). *J. Cell. Physiol.* **72**, *Suppl. 1*, 145.

99. WRIGHT, B. E. (1966). *Science* **153**, 830.

100. WRIGHT, B. E. and ANDERSON, M. L. (1960). *Biochim. et Biophys. Acta* **43**, 62.

101. WRIGHT, B. E. and ANDERSON, M. L. (1960). *Biochim. et Biophys. Acta* **43**, 67.

102. WRIGHT, B. E., BRUHMULLER, M., and WARD, C. (1964). *Dev. Biol.* **9**, 287.

103. WRIGHT, B. E. and DAHLBERG, D. (1967). *Biochemistry* **6**, 2074.

104. WRIGHT, B. E., DAHLBERG, D., and WARD, C. (1968). *Arch. Biochem. Biophys.* **124**, 380.

105. WU, R. and RACKER, E. (1963). In *Control Mechanisms in Respiration and Fermentation* (B. E. Wright, ed.). New York: The Ronald Press Company.

2

The Intermediary Metabolism of a Model System: the Cellular Slime Mold

"Metabolic chemistry is in its infancy. To paraphrase Newton's famous metaphor, we have amused ourselves with the shiny pebbles of metabolic sequences (and the smaller pebbles of individual enzymic reactions), while before us lay largely unperceived the ocean of interrelation and regulation. We do not yet really understand any of our pebbles, and we have only begun to notice a bit of salt spray in the air."

D. E. Atkinson, 1965 [4]

Life Cycle and General Metabolism

To describe the effects of metabolite flux in a differentiating system and to identify the variables which may be critical to the metabolic changes underlying differentiation represents a challenge that can be approached only by using the simplest of model systems. Hopefully, the conclusions reached will be applicable, at least in principle, to more complex organisms.

The cellular slime mold, *Dictyostelium discoideum*, is a microorganism exhibiting a primitive form of differentiation in which only two major cell types are involved: one type is transformed into the stalk, and the other into the spores of the final fruiting body [9]. We can begin a description of the life cycle at the stage of the vegetative amoebae, which grow indefinitely in the presence of sufficient food (Fig. 3). Under starvation conditions, however, (after the amoebae have been washed and transferred to nonnutrient agar) growth ceases, and differentiation is initiated. All the cells present at that

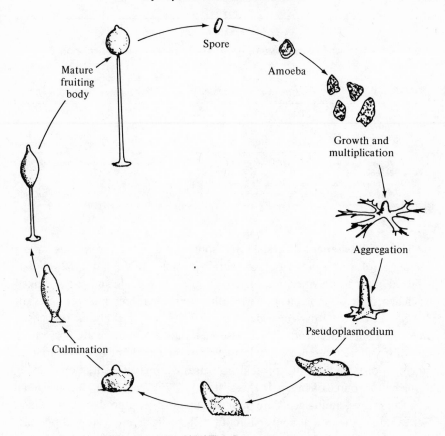

Figure 3. Life cycle of *Dictyostelium discoideum*.

time aggregate in response to cyclic AMP [35] to form a pseudoplasmodium composed of thousands of cells already partially differentiated into two types. The apical one-third of this multicellular body becomes the stalk, and the rear two-thirds the spores of a final fruiting body, the sorocarp. The latter is constructed during a process called culmination, initiated about 20 hours after the onset of starvation at 23° C. As might be expected for a system dependent upon the utilization of endogenous reserves, cell weight and volume decrease over the course of differentiation. For convenience in the type of analysis to be discussed in Chap. 3, these data are expressed in terms of packed cell volume (p. c. v.) and are summarized in Table 2. The rate of respiration also drops during differentiation and can be stimulated by exogenous glucose in the later stages when the normal endogenous energy sources are presumably limiting [38].

Table 2

DRY WEIGHT-CELL VOLUME DETERMINATIONS DURING DIFFERENTIATION[a]

Stage	Dry wt. (mg/ml)	p.c.v./ml	Corrected p.c.v./ml[c]	Dry wt./ml p.c.v.
Agg.	27.4	.195	.162	169
Culm.	23.3	.157	.106	220
Soro.[b]	18.2			

[a] These figures represent mean values from two experiments. Three determinations were made at each stage using identical cell aliquots.

[b] 15 hours from the time of culmination at 23°C.

[c] The per cent of the packed cell volume represented by cells is, at aggregation, 84.2 and at culmination, 66.7 [5]. The values at sorocarp were not included as they were highly variable and not required in any calculation.

The rigid material of the stalk and spore coats in the sorocarp is largely composed of an insoluble polysaccharide complex of cellulose [52] (a β-1, 4-linked glucose polymer) and glycogen [68] (an α-1, 4- and α-1, 6-linked branched glucose polymer). Two other carbohydrate end products of differentiation are a disaccharide, trehalose [13] (an α, α'-glycosyl glucose), and a mucopolysaccharide [65]. These materials are undetectable or, in the case of trehalose, present at very low levels at the beginning of differentiation (aggregation). Endogenous protein decreases during differentiation, serving as the major source of energy in this system [27, 76]. Very little gluconeogenesis occurs; thus, amino acids are poor precursors of end-product saccharides. For example, the rate of conversion of [^{14}C]-aspartate into cell wall polysaccharides during culmination is only a few percent of the rate of [^{14}C]-glucose incorporation [15, 17, 47]. A 50% inhibition by iodoacetate of the incorporation of [^{14}C]-aspartate into polysaccharide did not inhibit differentiation. An enzyme essential to gluconeogenesis, fructose-1, 6-diphosphatase, has unusually low activity and lacks regulatory properties associated with this enzyme as isolated from gluconeogenic systems [8, 16]. Also consistent with the utilization of protein as an energy source are the regulatory characteristics of phosphofructokinase, which suggest that this enzyme does not play a central role in energy-production from carbohydrates [7]. Soluble glycogen (as opposed to insoluble cell wall glycogen), which is present early in differentiation, appears to be the major source of the end-product saccharides which accumulate during sorocarp construction [16, 78, 82]. In fact, the extent to which soluble glycogen decreases late in differentiation can roughly account for the increase in end-product saccharides [82] (see Table 3). Total carbohydrate per cell stays essentially constant [57, 69]. For reasons of simplicity, then, and for analytical purposes, we shall treat this system as an interconversion

Table 3

SACCHARIDE CONCENTRATIONS EXPRESSED AS GLUCOSE
EQUIVALENTS/ML PACKED CELL VOLUME AT AGGREGATION

AGGREGATION

Carbohydrate	Percent agg. dry wt. (100 mg = 0.58 ml p.c.v.)	μmoles glucose equiv./100 mg	mM (÷ 0.58 ml p.c.v.)
Glycogen[a]	5.0	28.0	48.5
Trehalose[b]	1.0	5.5	9.5
Glucose	0.25	1.4	2.4
Unknown (based on total sugar)			~40

SOROCARP

Carbohydrate	Percent soro. dry wt.	Percent agg. dry wt.[c]	μmoles glucose equiv./100 mg	mM (÷ 0.58)
Glycogen	1.5	1.1	6.2	10.7
Cell wall complex	4.5	2.9	16.5	27.
Trehalose[b]	4.0	2.6	14.6	20.
Mucopolysaccharide	1.0	0.6	3.3	5.7
Unknown (based on total sugar)				~40

[a] *KOH-extracted; TCA extraction gives lower values [57].*
[b] *Literature values [13] from cells on rich media. Recent data from cells on nonnutrient agar give lower values at both stages of differentiation [57].*
[c] *Sorocarp dry wt./cell is 66% of that at aggregation.*

of saccharides and express all carbohydrate components in terms of glucose units. To avoid complicating the picture by expressing saccharide concentrations in terms of a parameter that changes with time (e.g., dry weight), concentration at all stages is expressed in terms of percent of the *original* dry weight, at aggregation, in glucose equivalents/ml p.c.v. All analyses are begun at this stage to insure that residual growth has ceased and that the metabolic reorganization involved in changing from an exogenous energy source to a dependence on endogenous metabolism has occurred. From the data of Table 2, and from values in the literature expressed as percent dry weight, approximate saccharide concentrations are given in Table 3, at two stages of differentiation: aggregation and sorocarp.

What is the order of events in the conversion of soluble glycogen to end-product saccharides? We have chosen aggregation as the beginning of the analysis, with $t = 0$ min when soluble glycogen is at its maximum level; at about $t = 900$ min, the end products have accumulated in the sorocarp. An intentionally over-simplified description of this conversion pro-

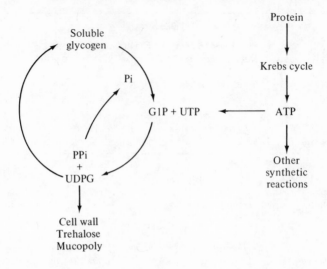

Figure 4.

cess is shown in Fig. 4. From the aggregation to the culmination stage of differentiation, the level of soluble glycogen remains fairly constant (the drop in level occurs between the culmination and sorocarp stages). During this early period, the "glycogen cycle" operates, and synthesis almost equals degradation (see below). However, this cycle speeds up about threefold between aggregation and culmination due in part to the enhanced activity of glycogen phosphorylase, an enzyme catalyzing the formation of G1P from glycogen, and increased levels of Pi, a substrate for this enzyme. By the time that culmination occurs, the rate of UDPG synthesis has reached a level adequate for the production of the observed amount of end-product saccharides over the required period of time [47]. The pyrophosphate (PPi) formed in this reaction appears to be the major source of Pi in this system— an active pyrophosphatase is present throughout differentiation [25]. During sorocarp formation a net decrease in glycogen occurs as its degradation continues and its rate of synthesis decreases. The breakdown of soluble glycogen may result in the release of bound glycogen synthetase, which then binds to an insoluble cell-wall primer for which it has a greater affinity [80]; there it catalyzes the formation of cell-wall glycogen [68]. Also at this time, an enzyme catalyzing the synthesis of trehalose is released from a masked form present from the beginning of differentiation [34, 62]. These and many other changes shift the fate of UDPG from soluble glycogen to the end products of differentiation. A few of the reactions involved may be seen in a more detailed description of the metabolic system (Fig. 5). "Survival" refers to the spore

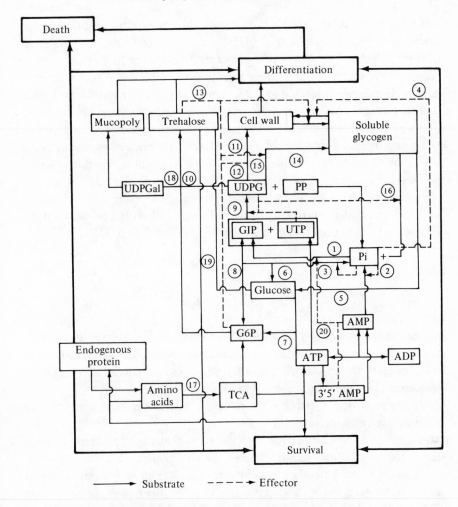

Figure 5. A partial metabolic map of reactions essential to differentiation in *D. discoideum*. Solid lines represent substrate-product relationships, and dashed lines indicate positive or negative modification of enzyme activity. "Death" refers to the stalk cells, and "Survival" to the spore cells, both of which are necessary to differentiation in this system.

cells, and "Death" to the stalk cells, which become vacuolated and critically depleted of reserve materials.

We are aware of many more details than those indicated, including the presence and role of yet other metabolites not shown. However, the purpose of this diagram is not to be right-up-to-date, but rather to illustrate the kind of complexity which *must* be taken into account in our analysis if we are to understand the critical variables underlying end-product accumulation in this system. All of the metabolites shown and the enzymes catalyzing the reactions indicated are essential to differentiation; our first problem is to discover which of these are primary critical variables. For example, is a change in substrate flux or enzyme activity responsible for affecting the rate of a specific reaction essential to the accumulation of an end product over the time period between aggregation ($t = 0$ min) and sorocarp construction ($t = 630$ min)? We shall see that the primary critical variable for the accumulation and increased rate of synthesis of UDPG is the increased rate of availability of G1P. Thus, secondary critical variables controlling this reaction become those factors responsible for the change in flux of G1P—e.g., the increased activity of glycogen phosphorylase and the availability of one of its substrates, Pi. By contrast, a change in enzyme activity is a primary critical variable controlling the rate of glycogen degradation (see p. 76).

The Role of Metabolites

Before dealing specifically with the role of a few selected metabolites known to be involved in the conversion of glycogen to end-product saccharides, it is useful to know the approximate concentration of such materials during differentiation. These data are presented in Table 4.

In the following discussion, the numbers in circles refer to a reaction or effect that may be located in Fig. 5.

Inorganic phosphate accumulates to a concentration of about 0.05 M during differentiation—tenfold higher than is present at aggregation. Pi concentration is apparently a critical variable, since an exogenous supply at endogenous levels has a striking positive effect on the *rate* of differentiation (Fig. 6). Agar plates either with or without Pi and with the same number of cells were incubated for a period of time such that only a small fraction of the ultimately possible mature sorocarps had formed on the control plates [36]. (With longer incubation periods the number of sorocarps is the same on all plates.) Pi has a number of known effects on metabolism in *D. discoideum*. A tenfold increase in its concentration would (1) enhance the activity of glycogen phosphorylase ① about twofold [33], (2) inhibit the activity of a 5'-nucleotidase ② to increasing extents during differentiation [26], (3) inhibit the activity ③ of an acid phosphatase [24] attacking G6P and G1P ⑥, and (4) inhibit trehalose, cell-wall, and glycogen synthesis ④ at the high levels of Pi present in sorocarps (see previous chapter, Table 1).

Table 4

METABOLITE CONCENTRATION PROFILES DURING DIFFERENTIATION

Metabolite	Extent and direction of change		Approx. max. conc. (mM)	Reference
	Agg. to culm.	Culm. to soro.		
1. Pi	up 2 ×	up 5 ×	50.0	26
2. Glucose	up 4–10 ×	down 10 ×	1.0	15, 77
3. G6P	up 3 ×	down 5 ×	0.1	15, 77
4. G1P	up 3 ×	down 5 ×	0.01	48
5. Glutamate	up 2 ×	up 5 ×	1.0	70
6. UDPG	up 3 ×	down 10 ×	0.1	47, 77
7. Fructose-6-P	up 7 ×		0.01	15
8. Fructose-di-P	up 2 ×		0.01	15
9. Dihydroxyacetone-P	up 2 ×		0.1	15
10. Glyceraldehyde-3-P	up 3 ×		0.01	15
11. 3-P-glycerate	—		0.05	15
12. 2-P-glycerate	up 4 ×		0.05	15
13. P-enolpyruvate	—		0.05	15
14. Pyruvate	up 10 ×		0.5	15
15. UTP	—		0.1	48
16. Cyclic AMP	down 100 ×		?	10

Glucose accumulates to a concentration of about 1 mM at culmination and probably arises through the action of amylase ⑤ [33] and through the combined action of glycogen phosphorylase ① and the acid phosphatase ⑥ [24]. Exogenous glucose has a positive effect on the rate of differentiation at a concentration of 10 mM (Fig. 7) [36]. Glucose also stimulates respiration, especially at the later stages of differentiation [38]. A more sensitive assay for the probable effects of endogenous glucose on metabolic pathways during differentiation involves the use of [^{14}C]-glucose labelled in the 1 or 6 position. At precumulation exogenous [^{14}C]-glucose levels as low as 0.1 mM affect the relative rates of [^{14}C]-CO_2 evolution from 1-and 6-[^{14}C] glucose [77]. Thus, the occurrence of changing hexose levels during differentiation must exert an influence on the pathways of hexose metabolism [28]. Since an active glucokinase ⑦ [6] and phosphoglucomutase ⑧ [16] are present, hexose phosphate and UDPG levels are undoubtedly raised in the presence of exogenous glucose. The enhanced rate of availability of G1P could be a critical variable with respect to the rate of synthesis of UDPG ⑨ [48], which in turn affects the rate of glycogen synthesis ⑭ [56].

G6P accumulates to a level of about 0.1 mM, through the combined action of a kinase ⑦ specific for glucose [6] and also through the action of

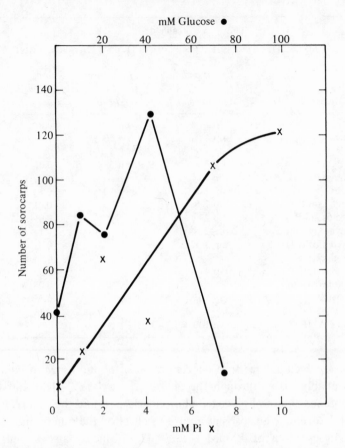

Figure 6. The effect of glucose and phosphate on the rate of differentiation in *D. discoideum*. Control and test plates had the same number of cells, and in time would produce the same number of sorocarps. To assess the effect of glucose or phosphate on the rate of differentiation, the number of sorocarps per plate were counted when only a few were present on control plates.

glycogen phosphorylase and phosphoglucomutase ⑧ [16]. G6P is a precursor ⑩ for trehalose, and since its endogenous levels limit the rate of the reaction [59, 72], G6P accumulation should ultimately stimulate trehalose synthesis. The accumulation of G6P early in differentiation would also stimulate ⑪ the rate of soluble glycogen synthesis ⑭ since it lowers the K_m of soluble glycogen synthetase for UDPG by a factor of ten [80]. The *decrease* in G6P late in culmination would reverse this effect, especially in view of the fact that at that time the enzyme is predominantly in the G6P-dependent form

[56]. The accumulation of cell wall glycogen ⑫ would be favored since the latter reaction is independent of G6P. Trehalose, formed from G6P and UDPG, in turn stimulates cell wall glycogen accumulation ⑬ by an unknown mechanism [72].

UDPG accumulates by a factor of three during differentiation to a concentration of about 0.1 mM at culmination; this level is known to limit the rate of synthesis ⑭ of soluble glycogen [56, 78], cell wall glycogen ⑮ [80], trehalose ⑩ [59], and UDP-galactose ⑱. Thus, the rate of all of these reactions should be influenced by a threefold increase in the level of UDPG. UDPG inhibits UDPG pyrophosphorylase ⑨ (see table 11, Chap 3) and glycogen phosphorylase in *D. discoideum* ⑯ [33]; the latter reaction could, therefore, be stimulated during early sorocarp construction as the UDPG pool level falls rapidly [77].

Glutamate oxidation ⑰ is an important process in a system dependent upon endogenous amino acids and proteins as an energy source [28]. The rate of this oxidation was measured *in vivo* ([^{14}C]-CO_2 formation from 1-[^{14}C] glutamate) and found to increase sevenfold during differentiation; the enzyme level did not change, but glutamate increased in concentration about tenfold, reaching levels of 1.0 mM [70]. Judging from the S/K_m ratio, the accumulation of substrate could account for the sevenfold increase in reaction rate. Hence, glutamate availability appears to be a critical variable in controlling the rate of this reaction. The same conclusion has been reached concerning the intermediates of the glycolytic pathway [16].

Although this list could be extended, these examples should suffice to indicate that changing levels and flux of endogenous metabolites during differentiation may exert a significant influence on the rate of substrate-limited reactions and on enzyme activities, as positive and negative modifiers. The absolute level of metabolites varies considerably from one group of cells to another [15, 75]; however, under the steady-state conditions of the living cell, we shall see that the important parameter concerns *changes* in metabolite flux and concentration [60]. Data on the relative concentration of metabolites are obtained by killing cells immediately with acid or heat and are therefore likely to be more reliable than comparable data on enzymes, which are intrinsically more labile and must be isolated in a manner permitting continued changes in their activity. Metabolite levels are usually obtained by harvesting the cells into acid as the initial step. It may be argued that, during the time and stress of this procedure, the steady-state pool levels could change. This would be particularly true of metabolites such as the nucleotides with rapid turnover rates relative to their steady-state levels. Nucleotide levels were therefore compared at two stages of differentiation by using both the

Table 5

Treatment   Nucleotide levels (mM)

	Pseudoplasmodium			Sorocarp		
	ATP	ADP	AMP	ATP	ADP	AMP
Perchloric Acid	0.93 ± 0.09	0.30 ± 0.03	0.10 ± 0.02	1.37 ± 0.40	0.42 ± 0.03	0.08 ± 0.01
Liquid Nitrogen	0.77 ± 0.12	0.30 ± 0.10	0.16 ± 0.07	1.45 ± 0.10	0.42 ± 0.05	0.09 ± 0.02

perchloric acid method and by killing the cells within a second by direct flooding with liquid nitrogen. The frozen cells were then transferred to perchloric acid [61]. As can be seen from Table 5, no significant differences were found.

The Mechanism of Changes in Metabolite Concentration

If a particular metabolite is identified as a primary critical variable, a change in its flux and/or concentration will affect the rate of a reaction essential to end-product accumulation; our attention is then turned to preceeding events critical to this change in flux and these events then become secondary critical variables and so on back in time until $t = 0$ min. A good example is the increased flux of G1P, which will be shown to be a primary critical variable in the next chapter. The source of G1P in this system is glycogen, and the enzymes involved are glycogen phosphorylase and/or, via a longer route, amylase, glucokinase, and phosphoglucomutase (see Fig. 5). The latter three enzymes apparently do not change significantly in activity, but both glycogen phosphorylase activity and Pi concentration do change in a manner that could account for the increased rate of G1P production from glycogen. This enzyme and substrate are, therefore, candidates for secondary critical variables. Tertiary critical variables for G1P accumulation would include mechanisms responsible for the enhanced enzyme activity (e.g., activation by cyclic AMP, etc.) and Pi accumulation (e.g., diminished oxidative phosphorylation, in turn due to a depletion of protein which is used as an energy source, and so on). As discussed in the previous chapter (see also Chap. 4), the flux of metabolites into this system will be largely controlled by "pacemaker" enzymes such as glycogen phosphorylase, trehalase, and protease, which occupy key positions in the utilization of the ultimate sources of substrates for anabolic and catabolic pathways. Changes in the activity of such enzymes are, therefore, implicated as primary critical variables in controlling the rate of utilization of their respective substrates.

The Role of Enzymes

A number of enzymes have been examined over the course of differentiation, and the *in vitro* activity profiles of some of them are presented in Tables 6 and 7. Enzymes of particular interest, of course, are those in the latter table, as they appear to undergo significant changes in activity and to be specifically involved in the accumulation of end products. Referring to the introduction, however, it is clear that such enzymes are no more essential to end-product accumulation than those in Table 6 that are, for example, concerned with amino acid catabolism or the tricarboxylic acid (TCA) cycle. However, metabolite flux in the latter general areas of metabolism appears to be substrate controlled; changes in the concentration of the enzymes involved are not criti-

Table 6

ENZYME ACTIVITY PROFILES DURING DIFFERENTIATION:
APPARENT CHANGES OF THREEFOLD OR LESS

Enzyme	Reference
1. Glucokinase	6, 16
2. Phosphoglucomutase	16
3. Phosphoglucose isomerase	16
4. Phosphofructokinase	7, 16
5. Fructose diphosphatase	8
6. Aldolase	16
7. Glyceraldehyde-3-P dehydrogenase	16, 71
8. Phosphoglycerate kinase	16
9. Phosphoglycerate mutase	16
10. Enolase	16
11. Pyruvate kinase	16
12. Lactate dehydrogenase	16, 71
13. Glucose-6-P dehydrogenase	71
14. Isocitrate dehydrogenase	71
15. Glutamate dehydrogenase	71
16. Alanine-α KG transaminase	71
17. β-glucosidase	18, 55
18. Pyrophosphorylase	25
19. α- and β-amylase	33
20. Tyrosine transaminase	51

Table 7

ENZYME ACTIVITY PROFILES DURING DIFFERENTIATION:
APPARENT CHANGES OF MORE THAN THREEFOLD

Enzyme	Reference
1. 5'–nucleotidase	26, 40
2. Acid phosphatase	24
3. Soluble glycogen synthetase	56, 78
4. Cell wall glycogen synthetase	68, 80
5. Cellulose synthetase	57
6. Cellulase	55
7. T6P synthetase	34, 59
8. Trehalase	12
9. UDPG pyrophosphorylase	3, 71
10. UDPGal transferase	65, 67
11. Glycogen phosphorylase	21, 33
12. Acetylglucosaminidase	39
13. Cyclic AMP phosphodiesterase	14, 46
14. α-manosidase	41
15. UDPGal–4–epimerase	67

cal variables [16, 70]. It is more likely that an enzyme concerned with cellulose synthesis, for example, would undergo significant changes critical to the rate of cellulose accumulation. We shall choose a few enzymes from Table 7 and discuss their possible role in end-product accumulation; again the numbers in circles refer to reactions, which may be found in Fig. 5:

UDPG Pyrophosphorylase ⑨ catalyzes the reaction:

$$G1P + UTP \longrightarrow UDPG + PP$$

The activity of this enzyme is essential as UDPG is a precursor of at least four saccharides present in the sorocarp. This enzyme reaches a peak specific activity in extracts prepared at culmination [3, 23, 48, 71] and *in vivo* analyses also indicate that the rate of UDPG synthesis has increased at this time [47]. The enhanced enzyme level could therefore be responsible for the increased level and rate of synthesis of UDPG.

Soluble glycogen synthetase ⑭ *and cell wall insoluble glycogen synthetase* 15 utilize UDPG as a precursor in the presence of a primer [68, 78]. These enzymes appear to be identical by a number of criteria [80]. During sorocarp construction, a fraction of this enzyme shifts from soluble glycogen as primer to insoluble cell wall material as primer for which it has a greater affinity. A reciprocal relationship exists between (1) the amount of cell wall glycogen and soluble glycogen, (2) the specific activity of the two enzyme fractions, and (3) the specific radioactivity of soluble and insoluble cell wall glycogen following a short exposure to [^{14}C]-glucose at various periods during sorocarp maturation [42]. This shift in enzyme localization could be a critical variable in the increased rate of cell wall glycogen synthesis and the decreased rate of soluble glycogen synthesis *in vivo.*

Trehalose-6-Phosphate (T6P) Synthetase ⑩ *catalyzes the reaction:*

$$UDPG + G6P \longrightarrow T6P + UDP$$

This enzyme reaches a peak activity in extracts prepared at preculmination [59], prior to trehalose accumulation. This increase in enzyme activity could therefore be primarily responsible for an increase in trehalose turnover as a prerequisite to its rapid accumulation during sorocarp construction.

Trehalase ⑲ degrades the disaccharide trehalose to two molecules of glucose, and presumably plays a major role in energy production in˷ the germination process [12.] Maximal activity occurs in amoebae extracts; during multicellular differentiation, the intracellular specific activity is highest at aggregation and decreases thereafter. Trehalase activity could be responsible for the lack of trehalose accumulation early in differentiation as well as contribute to the accumulation of glucose. The absence of trehalase activity during sorocarp construction could be critical to the accumulation of trehalose at that time.

Glycogen phosphorylase ① catalyzes the formation of G1P from glycogen in the presence of Pi and reaches a peak activity in extracts prepared at culmination [21, 33]. This enhanced enzyme activity could be a primary critical variable controlling glycogen degradation, which supplies precursor material for the synthesis of the end products of differentiation. Secondary critical variables may be the activation of this enzyme by 3′,5′-AMP and 5′-AMP ⑳ [56].

The Mechanism of Changes in Enzyme Specific Activity in vitro During Differentiation

Glycogen phosphorylase

Should changes in the activity of a particular enzyme prove to be a primary critical variable in affecting the rate of a reaction, it is of interest to pursue the secondary critical variables involved—i.e., the mechanisms by which the change in enzyme activity occurs. For example, the analysis in Chap. 3 will implicate glycogen phosphorylase as such an enzyme. Increased activity *in vitro* of glycogen phosphorylase over the course of differentiation could result from (1) *in vitro* artifacts created during the preparation and assay of the enzyme [71, 79], (2) enzyme conversion to a more active form, (3) allosteric modification, or release of inhibition of preformed enzyme, (4) unmasking (5) enhanced enzyme levels due to an increased rate of enzyme synthesis or, (6) decreased rate of enzyme degradation, or (7) a combination of mechanisms. Although experiments designed to obtain evidence for (1) have been negative [33], support has recently been obtained for (2) and (3). Cyclic AMP stimulates glycogen degradation *in vivo* (Fig. 7). Data suggest that 3'5'-AMP may mediate the conversion of glycogen phosphorylase to a more active form of the enzyme. 5'-AMP may stimulate glycogen degradation by overcoming the inhibition of glycogen phosphorylase by G1P [56]. This effect is shown in Fig. 8. To ascertain whether enzyme protein increases in concentration and, if so, through an increase in its rate of synthesis or decrease in its rate of degradation requires enzyme purification to homogeneity and a combination of isotope incorporation and immunochemical studies similar to

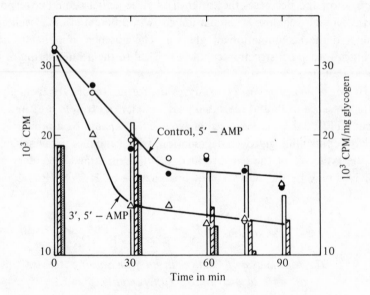

Figure 7. Changes *in vivo* with time in the total and specific radioactivity of glycogen. At 23 C, [^{14}C]-labeled cells in liquid cultures, pH 6.4, were exposed to a glucose chase (0.16 mM) in the presence of 3'5'-AMP or 5'-AMP (final concentration 1 mM). At the times indicated, total and specific radioactivity of glycogen was determined. The curves represent total counts per minute, and the block diagrams represent the specific radioactivity of glycogen in the control (open bar), in the presence of 5'-AMP (single cross-hatch), and in the presence of 3'5'-AMP (double cross-hatch). (The specific activity of glycogen in the presence of 3'5'-AMP at 90 min was 10,000 counts per min per mg of glycogen.) (See reference 56)

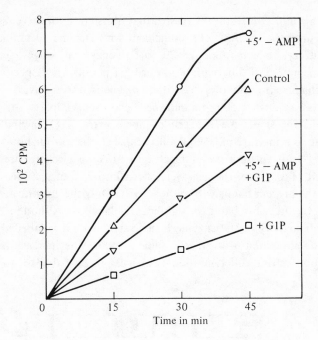

Figure 8. *In vitro* assay of glycogen phosphorylase. With
[14]C-labeled glycogen (1,620 counts per min per μ mole of
glucose equivalents) as substrate and a one-day-old dialyzed,
cell-free extract from pseudoplasmodia, glucose-1-phosphate
(GlP) was isolated and radioactivity was determined as
described in Materials and Methods. 5′-AMP and glucose-1-
phosphate were added to a final concentration of 1 mM in the
assay mixture which was incubated at 25 C for the time periods
indicated. (See reference 56)

those employed in mammalian systems [63]. Such studies are in progress for
UDPG pyrophosphorylase [28, 29].

UDPG pyrophosphorylase

During differentiation, the increase in specific activity of this enzyme is
correlated with an increase in enzyme protein [23]. The latter could be due
to an increased rate of enzyme synthesis or decreased rate of degradation.

During the growth phase of *D. discoideum*, glycogen and its precursor,
UDPG, are actively synthesized. Thus, UDPG pyrophosphorylase must be
synthesized at a rate such that it will double in amount at every cell division.
Upon starvation, there is every reason to believe that this enzyme will continue
to be synthesized, as it is also required during the differentiation phase of the

life cycle. Yet inhibitor studies suggested that translation for this enzyme does not occur until the twelfth hour, coincident with enzyme accumulation [2]. On the contrary, we have found [28, 29, 30] that enzyme is actively synthesized from the beginning of differentiation, during a period when specific activity does not change; presumably, therefore, enzyme is turning over.

Enzyme was purified 500-fold and specific antisera prepared for the selective precipitation of enzyme from crude amoeba extracts. Following exposure of amoebae to a mixture of [^{14}C]-amino acids for one to four hours, protein precipitated with antisera was subjected to SDS gel electrophoresis. After staining with fast green, the gels were sliced into 25 mm sections. Figure 9 correlates the protein bands with radioactivity. Only the enzyme area, represented by the heaviest band, is significantly labelled. Although the exact specific radioactivity of the enzyme is not yet known, it is labelled at a significant rate compared to average soluble protein. The increase in enzyme concentration during differentiation is at least in part due to a decreased

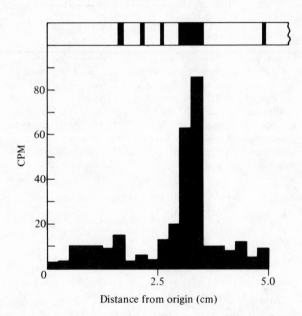

Distance from origin (cm)

Figure 9. Incorporation of [^{14}C] amino acids in UDPG pyrophosphorylase. Two-hour starved myxamoebae were incubated with 20μc [^{14}C] amino acid mixture (1 mc/mg) for 4 hours. Enzyme was isolated by precipitation with antisera and dissolved in 10 mM phosphate buffer containing 1% sodium dodecyl sulfate and 1% 2-mercaptoethanol. Following disc electrophoresis the protein bands in the gel were visualized by staining with fast green. Gel sections of 25 mm were dissolved in 0.5 ml 30% hydrogen peroxide and counted in a scintillation spectrometer in the presence of 10 ml Aquasol. A background count of 40 cpm was subtracted. Counting error was 10% [30].

rate of enzyme turnover, judging from comparative isotope incorporation studies at later stages of differentiation.

T6P synthetase

Enzyme activity is barely detectable in extracts of amoebae and reaches a peak in specific activity at preculmination [59]. However, by ammonium sulphate fractionation and heat treatment of amoebae extracts, enzyme activity is released from a masked state [34]. In fact, as indicated in Table 9, such treatment results in an increase of enzyme specific activity of some 700-fold, and the total number of enzyme units released per cell aliquot is comparable to that recovered from extracts prepared at preculmination. Evidence indicates that trehalase from *D. discoideum* is the inhibitor which masks T6P synthetase activity. The implications of this type of enzymatic regulation are exciting in view of its proposed role during differentiation [11]. The mechanism by which T6P synthetase becomes active in crude extracts prepared during successive stages of differentiation is called "*in vitro* unmasking". The mechanism by which this "*in vitro* unmasked" enzyme becomes active in the intact organism late in culmination is referred to as "*in vivo* unmasking", which will be discussed in Chap. 3.

Table 8
RECOVERABLE T6P SYNTHETASE ACTIVITY[a]

Fraction	Myxamoebae		Preculmination	
	Spec. act.	Units	Spec. act.	Units
Crude extract	0.13	4.9	38	1148
0–35% (NH$_4$)$_2$SO$_4$	30	47	46	153
0–35% (NH$_4$)$_2$SO$_4$ (10 min at 35° + 22 mM UDPG)	90	141	46	153
35–60% (NH$_4$)$_2$SO$_4$	23	141	32	223
Recoverable units		282		376

[a] *A unit of enzyme activity catalyzes the synthesis of 1 n mole of UDP per min at 25°*

The relevence of changes in enzyme specific activity to transcription and translation

In recent years, numerous studies have been carried out concerning the effect of actinomycin D on changes in enzyme specific activity during differentiation [3, 39, 41, 51, 59, 67]. Based upon these experiments, "maps" of periods of genetic transcription and mRNA translation for specific enzymes have been constructed. For two enzymes known to be present during vegetative growth (UDPG pyrophosphorylase and a 5-nucleotidase) transcriptional and translational events are postulated to occur more than four hours after the onset of differentiation. For another enzyme, also present during growth (N-acetyl glucosaminidase), transcription and translation are indicated as occurring at the beginning of differentiation. This is a curious situation, as inhibitor studies could give no information on which to base such a distinction—i.e., whether transcription continues as it exists in the vegetative amoebae or stops and is reinitiated after the onset of differentiation. These inhibitor studies have recently been extended to situations in which the rate of differentiation is altered by particular environmental conditions, also resulting in changes in the specific activity of various enzymes [20, 44, 45]. The latter changes are again interpreted to be controlled at the level of genetic transcription. Although a number of alternative interpretations exist [49, 71, 73, 81] the one offered by proponents of the inhibitor studies appears to be widely accepted. The clear implication of their interpretation is that sequential periods of genetic transcription during differentiation are primarily responsible for the control of metabolic events. Therefore, in our search for critical variables in this system, it is necessary to evaluate carefully the assumptions on which their interpretation is based:

1. It is assumed that an increase in enzyme specific activity represents an increase in the amount of enzyme protein—not, e.g., a conversion from preexisting, masked enzyme. Evidence for this assumption is now available for one enzyme, UDPG pyrophosphorylase [23]. However, it is clear that such evidence cannot be obtained for enzymes that are present early in differentiation but undetectable because they are masked (e.g., T6P synthetase (Table 8) and cellulase [55]), active in the synthesis of a product with different solubility characteristics from that assayed later in differentiation (e.g., cell-wall glycogen synthetase [80] and cellulose synthetase [57]), or present in a less active form (glycogen phosphorylase, see Figure 7).

2. It is assumed that an increase in the amount of enzyme protein results from an increase in the rate of enzyme synthesis. This assumption depends upon the validity of the first assumption and hence does not apply in the last five cases just cited. If the first assumption is valid, the second may be critically examined. The work of Schimke, discussed in Chap. 1, has clearly demonstrated that it is not always justifiable to equate enzyme accumulation with a change in the rate of enzyme synthesis. It is now known (see pp. 41-43) that UDPG

pyrophosphorylase is turning over prior to its accumulation. The latter is due at least in part to a decrease in the rate of enzyme degradation.

3. The third assumption underlying the interpretation of actinomycin D inhibition studies is that this drug, at levels above 100 μg/ml, specifically and selectively inhibits the formation of mRNA required in the synthesis of enzymes essential to differentiation. As will be seen, a number of other interpretations are possible.

As discussed in Chap. 1, if enzyme synthesis and degradation are inhibited to about the same extent, the level of an enzyme may remain unchanged, yet not indicate the presence of a stable mRNA template. In a case cited previously, tyrosine transaminase, actinomycin D at 5 μg/ml. inhibits degradation more than synthesis, resulting in an actual increase in enzyme level. A similar phenomenon appears to occur with this enzyme in the cellular slime mold [51]. It is significant that most slime mold enzyme levels "freeze", and do not change following the addition of actinomycin D, even at the time when the enzyme in the control cultures normally drops [51, 58]. Apparently, as Reel and Kenney have stressed [53] enzyme degradation as well as synthesis is prevented by these inhibitors. As in the case of tryptophan oxygenase, the accumulation of an enzyme such as UDPG pyrophosphorylase during differentiation could be due only to a decreased rate of degradation, resulting from the accumulation of stabilizing substrates or products (see pp. 41–43). Actinomycin D will act as an inhibitor regardless of whether this latter mechanism or changes in the rate of enzyme synthesis are responsible for enzyme accumulation. Therefore, the inhibition of an increase in enzyme concentration during differentiation does not necessarily indicate changes in the amount of an RNA template responsible for its synthesis. Such an inhibition is also consistent with a requirement for continuous protein synthesis and degradation. Moreover, at levels well below 100 μg/ml, this drug is known to have various nonspecific detrimental effects on metabolism [22, 32, 37, 50, 54, 64, 81]. Thus, actinomycin D may primarily prevent differentiation, and therefore also metabolic changes responsible for affecting enzyme activities (e.g., substrate accumulation, enzyme activation). The demonstration of "de novo" enzyme synthesis by the incorporation of [^{14}C]-amino acids or by the density-labeling method does not distinguish between changes in the rate of synthesis or degradation or both as the mechanism for enzyme accumulation. This point is frequently ignored or misunderstood [2].

In order to inhibit enzyme accumulation in the slime mold, actinomycin D must be used at concentrations which inhibit differentiation—i.e., higher than 100 μg/ml. As differentiation progresses, the system becomes less and less sensitive to actinomycin D—i.e., exposure of the cells to a given amount of the drug is followed by progressively longer and longer lag periods prior to the cessation of differentiation [49]. Thus, various correlations may be found between the time of addition of actinomycin D and its effect on enzyme accumulation, depending on such variables as the time at which the enzyme activity normally appears (by any one of a number of mechanisms), the rate of enzyme turnover, and cellular permeability to actinomycin D.

Assuming a specific action for this drug, the synthesis of the mRNA for an enzyme essential to energy metabolism may be very sensitive to inhibition. Subsequent interference with the synthesis of this enzyme may well prevent differentiation. However, it does not follow that this enzyme is unique to the differentiation process—only that it is one of the many prerequisites. Clearly, an enzyme detectable only in the later stages of differentiation may not appear if differentiation is prevented; regardless of the mechanism of enzyme appearance (see assumption 1 above), it would likely be dependent upon complex metabolic changes that accompany differentiation and do not occur in its absence. A trivial example concerns actinomycin D-sensitive enzyme changes of only two to threefold (e.g., tyrosine transaminase [51]). Protein and amino acid levels decrease by about 50% during differentiation, which will result in a twofold increase in specific activity for any enzyme not changing in concentration. A more interesting example is provided by T6P synthetase, which is undetectable early in differentiation not because it is absent, but because it is masked. Thus, although actinomycin D (at $>$ 100 μg/ml) will prevent the detection of this enzyme by stopping differentiation prior to the normal time of its appearance, any conclusions as to the *mechanism* responsible for its initial detection are unwarranted; deductions as to a period of genetic transcription are meaningless.

The effect of low levels of actinomycin D

At a concentration of 10 μg/ml, this drug completely prevents multicellular differentiation if added during growth or during the transition period between the growth and differentiation phases of the life cycle [49] (see Figure 10). Under these conditions RNA synthesis is inhibited about 50%. The period prior to aggregation is also most sensitive to γ-irradiation, suggesting active genetic transcription [19]. Net RNA levels are also decreasing at this time. If actinomycin D at 10 μg/ml is added at any time after the aggregation phase, however, subsequent multicellular differentiation is not prevented as shown in Fig. 10. As in the case of bacterial sporulation [1], such data may suggest that stable mRNA accumulates during growth prior to the occur-

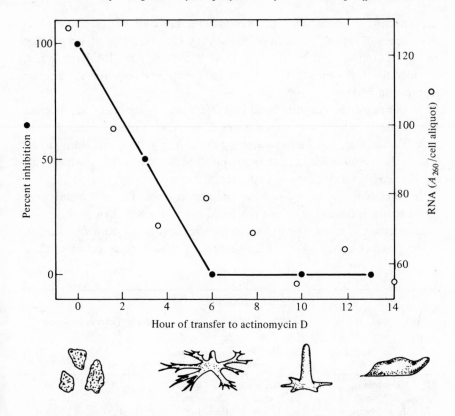

Figure 10. The correlation between RNA degradation in the absence of actinomycin D and inhibition of the rate of differentiation by actionmycin D at 10 μg/ml (for details, see text).

rence of aggregation and multicellular differentiation. This possibility was suggested as early as 1960 [49, 71, 74, 81], and has recently been reinforced by Mizukami and Iwabuchi [43], and by Hirschberg et al [31]. The latter investigators observed that, compared to growth of the amoebae, even the aggregation phase of differentiation is relatively insensitive to inhibition by actinomycin D: > 100-fold as much is required to block aggregation as to inhibit proliferation of the vegetative amoebae.

Recent studies of Deering et al [19], are of great interest in connection with the possibility that no new mRNA synthesis is necessary to the differentiation process following aggregation. A mutant of *D. discoideum*, apparently defective in its DNA repair system, is γ-irradiated to an extent preventing its duplication and cell proliferation. Yet these cells aggregate and proceed through normal differentiation, although the spores they form are incapable of further growth. These data are compatible with the possibility that all of the mRNA required for multicellular differentiation was present prior to these stages of the life cycle. In our hands, significant activity of any enzyme we have examined can be observed at very early stages of differentiation, provided that sufficient effort is made to find the optimal conditions of enzyme preparation and assay.

In summary, the enzymes listed in Table 9 are among those that increase or decrease in activity at various stages of differentiation. Changes in their activity, therefore, will be prevented by the earlier addition of actinomycin D at levels inhibiting further differentiation. The mechanism by which enzyme specific activity changes is also indicated, if known.

In conclusion, we have now described the presence of several metabolites and enzymes in the cellular slime mold and have discussed their possible role as substrates, modifiers, or catalysts in specific biosynthetic pathways leading to end-product accumulation. Many more could have been included. Con-

---

Table 9

MECHANISM OF CHANGES IN ENZYME SPECIFIC ACTIVITY DURING DIFFERENTIATION

Enzyme	Direction of change	Mechanism of change	References
1) 5'Nucleotidase	Increase	Unknown	26, 40
2) Acid phosphatase	Increase	Unknown	24
3) Soluble glycogen synthetase	Decrease	Partial relocation during culmination to insoluble cell wall material	56, 78
4) Glycogen phosphorylase	Increase	In part due to conversion from inactive form by 3'5'AMP	21, 33, 56
5) Trehalase	Decrease	Excreted	12
6) Acetylglucosaminidase	Increase	Unknown	39
7) Cellulase	Increase	In part due to unmasking	55
8) T6P synthetase	Increase	Entirely due to unmasking	34, 59, 62
9) Cell wall glycogen synthetase	Increase	Partial relocation during culmination to insoluble cell wall material	68, 80
10) Cell wall cellulose synthetase	Increase	Initially forms an alkali soluble product (?)	57
11) UDPGal transferase	Increase	Unknown	67
	Decrease	Excreted	66
12) UDPG pyrophosphorylase	Increase	Alteration in enzyme turnover	23, 28, 29
13) UDPGal epimerase	Increase	Unknown	67
	Decrease	Excreted	67
14) α-Mannosidase	Increase	Unknown	41
15) Extracellular Cyclic AMP phosphodiesterase	Decrease	Protein inhibitor produced at aggregation	14, 46

sidering only those selected, their multiple effects and interdependence are so complicated, and the range of possible control mechanisms so complex, that it is difficult to distinguish random variation from significant change. Many essential metabolites and enzymes undergo changes in concentration or activity as a function of time over the course of differentiation. Mechanistically, the changes of interest are those which are primary critical variables —i.e., directly affect the rate of a reaction essential to end-product accumulation in the intact cell subsequent to $t = 0$ min (aggregation). As we shall see, it has proven exceedingly difficult to interpret the relevence of *in vitro* data, particularly that pertaining to enzymes, to metabolism in the living organism. Consideration of such data alone are not only inadequate but often misleading in our attempts to decide which cellular components actually function as critical variables. Our analysis can be enormously strengthened by two kinds of additional information: (1) data on the rate of reactions *in vivo*, and (2) a kinetic model of the system. The next chapter attempts to illustrate how useful such a model can be in reducing the conceptual difficulties inherent in the analysis of a complex system, and in stimulating us to think and ask questions in terms of the steady state conditions of the living cell. Specifically, a kinetic model can help to organize and unify a mass of data, reveal inconsistencies and contradictions in our ideas about our data, serve as a framework in which to assess the relevence of *in vitro* analyses to metabolism *in vivo*, and finally, reveal relationships and objectives not previously apparent.

REFERENCES

1. ARONSON, A. I. and DEL VALLE, M. R. (1964). *Biochim. et Biophys. Acta* **87,** 267.

2. ASHWORTH, J. M. (1971). Symp. Soc. Exp. Biol. No. XXV, 27.

3. ASHWORTH, J. M. and SUSSMAN, M. (1967). *J. Biol. Chem.* **242,** 1696.

4. ATKINSON, D. E. (1965). *Science* **150,** 851.

5. BARRAVECCHIO, J., BAUMANN, P., and WRIGHT, B. E. (1969). *J. Appl. Microbiol.* **17,** 641.

6. BAUMANN, P. (1969). *Biochemistry* **8,** 5011.

7. BAUMANN, P. and WRIGHT, B. E. (1968). *Biochemistry* **7,** 3653.

8. BAUMANN, P. and WRIGHT, B. E. (1969). *Biochemistry* **8,** 1655.

9. BONNER, J. T. (1967). *The Cellular Slime Molds.* Princeton, New Jersey: Princeton University Press.

10. BONNER, J. T., BARKLEY, D. S., HALL, E. M., KONIJN, T. M., MASON, J. W., O'KEEFE, G., and WOLFE, P. B. (1969). *Develop. Biol.* **20**, 72.

11. CALVO J. M. and FINK, G. R. (1971). *Ann. Rev. Biochem.* **40**, 943 (E. E. Snell, ed.)

12. CECCARINI, C. (1967). *Biochim. et Biophys. Acta* **148**, 114.

13. CECCARINI, C. and FILOSA, M. (1965). *J. Cell Comp. Physiol.* **66**, 135.

14. CHANG, Y. (1968). *Science* **160**, 57.

15. CLELAND, S. V. (1969). Ph.D. Dissertation "Gluconeogenesis and Glycolysis in *Dictyostelium discoideum*," Northwestern University.

16. CLELAND, S. V. and COE, E. L. (1968). *Biochim. et Biophys. Acta* **156**, 44.

17. CLELAND, S. V. and COE, E. L. (1969). *Biochim. et Biophys. Acta* **192**, 446.

18. COSTON, M. B. and LOOMIS, W. F., JR. (1969). *J. Bact.* **100**, 1208.

19. DEERING, R. A., ADOLF, A. C., and SILVER, S. M. (1972). *Int. J. Rad. Biol.* **21**, 235.

20. ELLINGSON, J. S., TELSER, A. and SUSSMAN, M. (1971) *Biochim. et Biophys. Acta* **244**, 388.

21. FIRTEL, R. A. and BONNER, J. (1970). *Fed. Proc.* **29**, 669.

22. FLICKINGER, C. J. (1971). *J. Cell Biol.* **49**, 221.

23. FRANKE, J. and SUSSMAN, M. (1971). *J. Biol. Chem.* **246**, 6381.

24. GEZELIUS, K. (1966). *Physiol. Plant.* **19**, 946.

25. GEZELIUS, K. (1968). *Physiol. Plant.* **21**, 35.

26. GEZELIUS, K. and WRIGHT, B. E. (1965). *J. Gen. Microbiology* **38**, 309.

27. GREGG, J. H., HACKNEY, A. L., and KRIVANEK, J. O. (1954). *Biol. Bull.* **107**, 226.

28. GUSTAFSON, G. and WRIGHT, B. E. (1972). *CRC Critical Reviews in Microbiology* **1** (4) (A. I. Laskin and H. Lechevalier, eds.). Ohio: CRC Press, p. 453.

29. GUSTAFSON, G. and WRIGHT, B. E. (1971). *Fed. Proc.* **30**, 1069 Abs.

30. GUSTAFSON, G. and WRIGHT, B. E. (unpublished data).

31. HIRSCHBERG, E., CECCARINI, C., OSNOS, M., and CARCHMAN, R. (1968). *Proc. Nat. Acad. Sci. U. S.* **61**, 316.

32. HONIG, G. R. and RABINOWITZ, M. (1965). *Science* **149**, 1504.

33. JONES, T. H. D. and WRIGHT, B. E. (1970). *Fed. Proc.* **29**, 670.

34. KILLICK, K. and WRIGHT, B. E. (1972). *J. Biol. Chem.* **247**, 2967.

35. KONIJN, T. M., BARKLEY, D. S., CHANG, Y. Y., and BONNER, J. T. (1968). *The Am. Natur.* **102**, 225.

36. KRICHEVSKY, M. I. and WRIGHT, B. E. (1963). *J. Gen. Microbiol.* **32**, 195.

37. LASZIO, J., MILLER, D. A., McCARTHY, K. S., and HOCHSTEIN, P. (1966). *Science* **151,** 1007.

38. LIDDEL, G. U., and WRIGHT, B. E. (1961). *Develop. Biol.* **3,** 265.

39. LOOMIS, W. F., JR. (1969). *J. Bact.* **97,** 1149.

40. LOOMIS, W. F., JR. (1969). *J. Bact.* **100,** 417.

41. LOOMIS, W. F., JR. (1970). *J. Bact.* **103,** 375.

42. MARSHALL, R. and WRIGHT, B. E. (unpublished data).

43. MIZUKAMI, Y. and IWABUCHI, M. (1970). *Exptl. Cell Res.* **63,** 317.

44. NEWELL, P. C., FRANKE, J., and SUSSMAN, M. (1972). *J. Mol. Biol.* **63,** 373.

45. NEWELL, P. C. and SUSSMAN, M. (1970). *J. Mol. Biol.* **49,** 627.

46. PANNBACKER, R. G. (1970). *Bacteriol. Proc., 70th Annual Meeting*, p. 23.

47. PANNBACKER, R. G. (1967). *Biochemistry* **6,** 1283.

48. PANNBACKER, R. G. (1967). *Biochemistry* **6,** 1287.

49. PANNBACKER, R. G. and WRIGHT, B. E. (1966). *Biochem. Biophys. Res. Commun.* **24,** 334.

50. PASTAN, I. and FRIEDMAN, R. M. (1968). *Science* **160,** 316.

51. PONG, S. S. and LOOMIS, W. F., JR. (1971). *J. Biol. Chem.* **246,** 4412.

52. RAPER, K. B. and FENNELL, D. I. (1952). *Bull Torrey Botan. Club.* **79,** 25.

53. REEL, J. R. and KENNEY, F. T. (1968). *Proc. Nat. Acad. Sci. U. S.* **61,** 200.

54. REVEL, M., HIATT, H. H., and REVEL, J. (1964). *Science* **146,** 1311.

55. ROSNESS, P. A. (1968). *J. Bact.* **96,** 639.

56. ROSNESS, P. A., GUSTAFSON, G., and WRIGHT, B. E. (1971). *J. Bact.* **108,** 1329.

57. ROSNESS, P. A. and WRIGHT, B. E. (unpublished data).

58. ROTH, R., ASHWORTH, J. M., and SUSSMAN, M. (1968). *Proc. Nat. Acad. Sci. U. S.* **59,** 1235.

59. ROTH, R. and SUSSMAN, M. (1968). *J. Biol. Chem.* **243,** 5081.

60. ROTHMAN, L. B. and CABIB, E. (1969). *Biochemistry* **8,** 3332.

61. RUTHERFORD, C. and WRIGHT, B. E. (1971). *J. Bact.* **108,** 269.

62. SARGENT, D. and WRIGHT, B. E. (1971) *J. Biol. Chem.* **246,** 5340.

63. SCHIMKE, R. T. (1969). In *Current Topics in Cellular Regulation* 1 (B. L. Horecker and E. R. Stadtman, eds.). New York: Academic Press Inc., p. 77.

64. SOEIRO, R. and AMOS, H. (1966). *Biochim. et Biophys. Acta* **129,** 406.

65. SUSSMAN, M. and OSBORN, M. J. (1964). *Proc. Nat. Acad. Sci. U. S.* **52,** 81.

66. SUSSMAN, M., and SUSSMAN, R. (1969). *Symp. Soc. Gen. Microbiol.* **19,** 403.

67. TELSER, A. and SUSSMAN, M. (1971). *J. Biol. Chem.* **246,** 2252.

68. WARD, C., and WRIGHT, B. E. (1965). *Biochemistry* **4**, 2021.

69. WHITE, G. J. and SUSSMAN, M. (1961). *Biochim. et Biophys. Acta* **53**, 285.

70. WRIGHT, B. E. (1963). *Bact. Rev.* **27**, 273.

71. WRIGHT, B. E. (1964). In *Biochemistry and Physiology of Protozoa* **3**. (*S. H. Hutner*, ed.). New York: Academic Press, Inc., p. 341.

72. WRIGHT, B. E. (1965). In *Developmental and Metabolic Control Mechanisms and Neoplasia* (*D. N. Ward*, ed.). Baltimore, Md: The Williams and Wilkins Co., p. 296.

73. WRIGHT, B. E. (1968). *J. Gen. Physiol. Supp.* 1, **72**, 145.

74. WRIGHT, B. E. (1960). *Proc. Nat. Acad. Sci., U. S.* **46**, 798.

75. WRIGHT, B. E. (1966). *Science* **153**, 830.

76. WRIGHT, B. E. and ANDERSON, M. L. (1960). *Biochim. et Biophys. Acta* **43**, 62.

77. WRIGHT, B. E., BRUHMULLER, M., and WARD, C., (1964). *Dev. Biol.* **9**, 287.

78. WRIGHT, B. E. and DAHLBERG, D. (1967). *Biochemistry* **6**, 2074.

79. WRIGHT, B. E. and DAHLBERG, D. (1968). *J. Bact.* **95**, 983.

80. WRIGHT, B. E., DAHLBERG, D., and WARD, C. (1968). *Arch. Biochem. Biophys.* **124**, 380.

81. WRIGHT, B. E. and PANNBACKER, R. G. (1967). *J. Bact.* **93**, 1762.

82. WRIGHT, B. E., SIMON, W., and WALSH, B. T. (1968). *Proc. Nat. Acad. Sci. U. S.* **60**, 644.

3

Using Kinetic Models to Find
the Critical Variables

*". . . it is necessary to uncover, both experimentally and logically
the causal connections of a system without isolating the steps of which it is composed.
The language in which such a system is described must of necessity be
the language of molecular interactions, namely kinetics. Our conventional logical
apparatus, which is essentially a linear one and lacks quantitative rigour, cannot handle
most of the situations which are of the essence of interacting systems.
Some of the conclusions of the treatment which follows may therefore appear
intuitively strange—but so much the worse for intuition"*

H. Kascer, 1963 [16]

Introduction

Models by definition simplify a system—hopefully in such a way as to be essentially correct and to ignore complexities of lesser importance. By epitomizing the system, insight may be gained into the nature of variables most critical to its function; parameters lacking particular bearing on function may also be revealed. Oversimplified and inaccurate though it may be, explicit model-making exposes the role of, or need for, data of previously unsuspected importance: ". . . progress emerges from error far more easily than from chaos [12]." Even simple models is such as the ones to be described have been enormously helpful in analyzing this differentiating system. The mind cannot retain and simultaneously integrate the entire complex of data involved—it almost seems necessary to ignore one aspect of metabolism in order to concentrate on another. The computer has no such difficulties: Time and again the model produces an unexpected answer that only then becomes obvious—along with the previously overlooked data or relationship.

The first model to be described depicts only the overall stoichiometry of the saccharide interconversion occurring during differentiation; it assumes simple initial velocity kinetics and ignores the role of known effectors and end-product inhibitors. The second, expanded model includes more specific reactions, several sophisticated and accurate enzyme kinetic expressions, and cases of end-product inhibition; we are presently incorporating the role of known enzyme modifiers such as G6P in the case of glycogen synthetase. A continual interplay between model exploration, predictions, and experimental findings is the basis for expansion and refinement of the kinetic model. Our ultimate aim is to simulate quantitatively all of the metabolic pathways critical to differentiation in this simple system. However, we must proceed with care. The compatability of a model with the available data by no means demonstrates the correctness of the model, which is primarily valuable as a guideline for further research. A model can be made to perform under conditions differing from those directly involved in its development and thus make specific predictions: The only real test of the validity of a model lies in its predictive value.

We have seen in the previous chapter that enzyme specific activity may change during differentiation by a number of mechanisms. Regardless of the mechanism, the really important question which concerns all of us is the relevence of such changes to the control of metabolism in the intact cell. Fortunately, this basic problem can be tackled, albeit indirectly, through the use of kinetic models. In fact, it is difficult to imagine an alternative technique for gaining comparable insight into the metabolic complexities of the differentiating cell. Such models provide a framework for organizing data and for probing the interaction of enzyme activities with metabolite accumulation patterns and flux values determined *in vivo*. The role that each parameter plays in controlling metabolic events during differentiation is assessed not only in terms of its individual kinetic properties, but also in terms of its kinetic position in the overall metabolic system. In short, an attempt is made to simulate the dynamics of metabolism under the conditions prevailing in the differentiating organism. Apparently this is by no means a hopeless task, for many predictions made by such models have now been substantiated experimentally.

In the differentiating cell, the activity of an enzyme may be affected by any one of a multitude of known and unknown factors, which in the last analysis are very difficult to assess. These include compartmentalization of enzyme from substrate or end-product, masking, and positive or negative modulation of enzyme activity by metabolites or macromolecules. Enzymes are often isolated in a manner allowing changes in their activity and must be assayed under nonphysiological conditions. It is, therefore, highly desirable to have criteria other than the measurement of enzyme specific activity in cell extracts by which to assess enzyme activity in the intact cell. A more sophisticated and accurate estimation of enzyme activity may be attempted within the framework of a kinetic model of the metabolic system under analysis. Enzyme activities are estimated *indirectly* through an integration of the relationship of their kinetic constants to cellular levels of substrates, end-product inhibitors and effectors, and through a knowledge of key reaction rates determined *in vivo*.

A specialized program called METASIM was developed to act as a general purpose metabolic simulator of *in vivo*, multienzyme systems undergoing long-term changes in metabolite concentrations and enzyme activities. METASIM is written in the language of the biochemist and can be used without a knowledge of programming or the details of differential equations. The principal features of this program are described in the appendix.

In order to proceed with this kind of analysis, simplifications and assumptions must, of course, be made. Metabolite accumulation patterns are assumed to be valid *in vivo*. In several cases already analyzed, compartmentalization does not appear to be metabolically significant [20, 25], but can be assessed when necessary [21]. In general, the important parameter concerns *changes* in metabolite concentration and flux. Flux values obtained *in vivo* are also considered to be valid; these are comparable whether determined under liquid conditions or during differentiation in the undisturbed state [17, 24]. It is assumed that kinetic enzyme mechanisms and constants (e.g., K_m and K_i) determined *in vitro* apply *in vivo* and do not change over the course of differentiation. However, the activity of an enzyme *in vivo* (equivalent to V_{max}), and the pattern of change in this activity during differentiation, are taken as unknowns in our analysis. These parameters are calculated, or predicted, based upon parameters assumed to be valid (e.g., *in vivo* flux values).

The rate expressions used vary in complexity according to the reaction and to our experimental knowledge of the reaction. For example, a nonreversible unimolecular reaction without significant product inhibition can be represented by an initial velocity expression, where V_1 is the maximum forward velocity, A the substrate concentration, and K_a the Michaelis binding constant.

$$\text{reaction rate} = V = \frac{V_1 A}{K_a + A}$$

The Michaelis binding constant K_a would be determined by *in vitro* measurements. To emphasize the fact that the maximum forward velocity

V_1 is determined not by an *in vitro* measurement, but is calculated from kinetic constants, *in vivo* measurements of the reaction rate, and examination of metabolite accumulation patterns with time, it is denoted by $'V_v.*$ Furthermore, in a differentiating system some enzymes undergo activation (or deactivation). This can be represented by making V_v a function of time, so that the rate expression becomes:

$$V = V_v(t) \cdot \frac{A}{K_a + A}$$

The computer program allows the activity of each enzyme to be arbitrarily modulated by a time function supplied by the user. We call $V_v(t)$ an activation function. This representation tells us nothing about the mechanism underlying the change in $V_v(t)$ as a function of time. It only means that there is some mechanism not explicitly included in the model which affects the enzymatic activity during differentiation. The existence of an activation function may, but does not necessarily imply that the enzyme is changing its concentration as a function of time. There may be masking, unmasking, or compartmentalization of the enzyme from its substrate. An activation function may even be invoked if an incorrect or incomplete kinetic mechanism is used to represent the reaction. When our estimated value of V_v at a particular point in time corresponds to our *in vitro* (V_{max}) value, perhaps we have correctly assumed or measured most of the parameters which are relevant *in vivo*. A lack of correspondence may suggest the presence, for example, of an unsuspected inhibitor and send us on a new search. At present, the principal use of the program in our laboratory is to provide a quantitative method for separating and analyzing the effects of changes in enzyme activity from those effects brought about by variations in the levels and flux of reactants—e.g., substrates and end-product inhibitors. The interrelationship of these types of variables is complex, and any analysis that cannot handle them quantitatively is hopelessly oversimplified.

Data Used in the Simplified Model

Two kinds of information were necessary to construct our first kinetic model of the conversion of soluble glycogen to the end products of differentiation in *D. discoideum*: (1) the concentration profiles of the relevant carbohydrates and intermediates involved (those shown in Fig. 11), and (2) the rate *in vivo* of UDPG synthesis (V_{UDPG}) at aggregation and culmination [35].

*This value was indicated by V_{max} in previous publications.

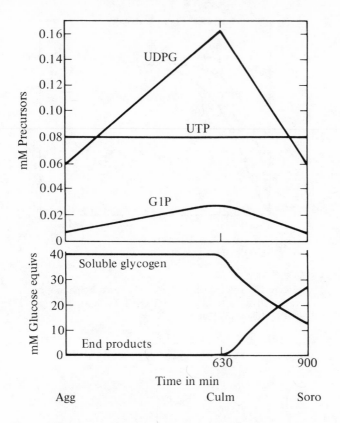

Figure 11. A schematic representation of the observed accumulation patterns of UDPG, G1P, UTP, soluble glycogen, and the various end-product saccharides that have been determined during differentiation. See Tables 3 and 4 of Chap. 2.

V_{UDPG} was determined by exposing intact cells to [14C]-uracil and taking samples every minute for the isolation of UTP and UDPG in order to follow the incorporation of isotope from precursor into product (Fig. 12).

V_{UDPG}, expressed as μmoles/min/ml packed cells, was found to increase about threefold from aggregation to culmination. Knowing the cellular concentrations of UTP and G1P, and the K_m's of the pyrophosphorylase for these two substrates, it was possible to calculate enzyme activity *in vivo* (V_v) from an initial velocity expression,

$$V_{UDPG} = V_v \left(\frac{1}{\dfrac{K_{UTP}}{[UTP]} + 1} \right) \left(\frac{1}{\dfrac{K_{G1P}}{[G1P]} + 1} \right)$$

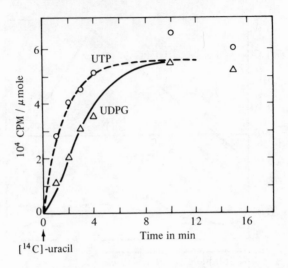

Figure 12. The incorporation *in vivo*, at culmination, of radioactivity into UTP and UDPG during incubation with [14C]-uracil [20].

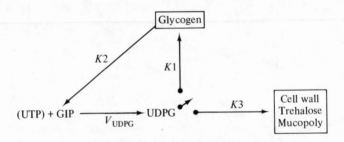

Figure 13. A simplified kinetic model of carbohydrate metabolism in *D. discoideum* [30, 35].

This value for V_v was used in the model shown in Fig. 13 (reminiscent of Fig. 4 in the previous chapter). $K1$, $K2$ and $K3$ are assumed rate constants which will be explained below. The end products are treated as a single component; the small arrow indicates a switch in the fate of UDPG and will be discussed later.

The following assumptions are implicit in this model:

1. The K_m values for G1P and UTP determined *in vitro* apply *in vivo* and do not change from aggregation to sorocarp.

2. The rate of synthesis of UDPG can be described by an initial velocity expression. Binary complexes have been assumed.

3. The rates of synthesis of soluble glycogen and end-product saccharides are directly proportional to the concentration of UDPG, that is, K_{UDPG} (the Michaelis constant for UDPG) is much larger than the concentration of UDPG. This is known to be the case in the synthesis of soluble glycogen [31], cell wall glycogen [28], and trehalose [22].

4. The rate of production of G1P is directly proportional to the concentration of soluble glycogen.

5. All the reactions are essentially irreversible under the conditions existing within the cell (e.g., an active pyrophosphorylase present throughout differentiation [10] pulls the reaction catalyzed by UDPG pyrophosphorylase in the direction of synthesis).

6. The end product saccharides are not degraded in the sorocarp.

7. There is no gluconeogenesis.

The simple differential equations describing the process are as follows:

$$\frac{d[\text{UDPG}]}{dt} = V_{\text{UDPG}} - K1\,[\text{UDPG}] \text{ before switchover}$$

$$\frac{d[\text{UDPG}]}{dt} = V_{\text{UDPG}} - K3\,[\text{UDPG}] \text{ after switchover}$$

$$\frac{d[\text{sol. glyc.}]}{dt} = K1\,[\text{UDPG}] - K2\,[\text{sol. glyc.}]$$

$$\frac{d[\text{end products}]}{dt} = K3\,[\text{UDPG}]$$

$$\frac{d[\text{G1P}]}{dt} = K2\,[\text{sol. glyc.}] - V_{\text{UDPG}}$$

$$\frac{d[\text{UTP}]}{dt} = 0$$

These equations were solved by incremental methods with a computer and the concentrations of the various species were plotted as a function of time (900 min from aggregation to sorocarp) [35]. Values for $K1$ and $K2$ were found by assuming that at aggregation no end-product saccharides are synthesized ($K3 = 0$) and that a steady-state exists; that is, the rate of UDPG synthesis equals the rate of soluble glycogen synthesis equals the rate of G1P production. [A slight net increase in degradation must occur between aggregation and eulmination to account for the accumulation of G1P and UDPG.] Since the rate of UDPG synthesis has been measured and since the concentrations of the appropriate species are all known, $K1$ and $K2$ can be determined. The value and variation in $K3$ after aggregation will be discussed later.

The features of the metabolic processes with which the model must be consistent are the variations as a function of time in the concentrations of UTP, G1P, UDPG, soluble glycogen, and the end-product saccharides as well as the increase *in vivo* in the rate of UDPG synthesis. The importance of various factors in producing these features may be assessed by varying the reaction parameters of the model and determining the resultant changes by use of the computer.

Figures 14A and 14B show the steady-state situation in which none of the reaction parameters are changed over the time scale of 900 min. The difference in ordinate scales in these two figures should be noted; the concentrations of UTP, G1P, and UDPG are negligible compared to that of soluble glycogen and are therefore hidden in the baseline of Fig. 14B. Figure 14C illustrates the effect on metabolite concentrations of tripling the V_v of UDPG pyrophosphorylase (equivalent to tripling the amount of active enzyme) in a linear fashion from 10 μmole/min/ml at aggregation to 30 μmole/min/ml at culmination. The only significant change is a threefold reduction of G1P from aggregation to culmination. There is no resultant increase in either the concentration of UDPG (Fig. 13C) or the rate (not indicated) of its synthesis; for while V_v has gone up by a factor of three, G1P has fallen by a factor of three. The reduction of G1P is absorbed by an insignificant increase in soluble glycogen; thus, the variation in the concentration of soluble glycogen with time is essentially identical to that in Fig. 14B.

Figure 14D demonstrates the results of reducing the V_v of the pyrophosphorylase by a factor of three linearly with time from aggregation to culmination. Once again the only effect is on G1P, which accumulates. For reasons similar to those discussed above, there is neither a decrease in the rate of UDPG synthesis nor a significant change in the concentration of soluble glycogen. Thus, alterations in the amount or activity of UDPG pyrophosphorylase affect only the concentration of G1P. In like fashion variations in $K1$ change only the concentration of UDPG.

If the rate constant $K2$ (representing glycogen phosphorylase) is tripled in a linear fashion from 0.0008 min^{-1} at aggregation to 0.0024 min^{-1} at culmination (i.e., given a threefold activation function), the results illustrated in Fig. 14E are obtained. The concentrations of both G1P and UDPG rise threefold, as does the rate of UDPG synthesis; this is in accord with the experimental data. However, there is no significant decrease in soluble glycogen, G1P, or UDPG from culmination to sorocarp, nor can any end-product saccharides accumulate. These latter features, observed *in vivo* (see Fig. 11), may be achieved with the model by setting $K1 = 0$ and $K3 = 0.5$ min^{-1} (the value previously held by $K1$) at culmination and maintaining these values until sorocarp (900 min). In reality, of course, such a "switch" is not instantaneous; it will be refined in future models. Changes in the assumed

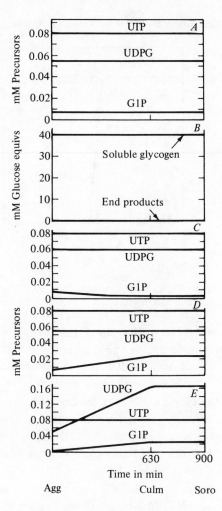

Figure 14 A and B. Computer outputs of the steady-state situation in which no parameter is varied over the time scale of 900 min. The time scale has been modified [14] to conform to later analyses; no previous results are affected [30].

C. Computer output resulting from a tripling of the amount of UDPG pyrophosphorylase V_v in a linear fashion from 10 μmole/min/ml at aggregation to 30 μmoles/min/ml at culmination.

D. Computer output resulting from a reduction of V_v by a factor of three linearly with time from aggregation to culmination.

E. Computer output resulting from a tripling of $K2$ in a linear fashion from 0.0008 min^{-1} at aggregation to 0.0024 min^{-1} at culmination [35].

61

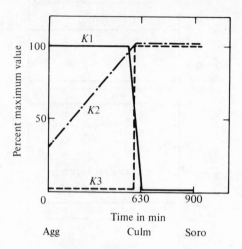

Figure 15. The assumed activation functions for the kinetic constants.

values of $K1$, $K2$ and $K3$ as a function of time are shown schematically in Fig. 15. If this pattern of parameter variation is added to that used in producing Fig. 14E, the changes illustrated in Fig. 11 are obtained, and essentially all the experimental observations are reproduced. Glucose units are "shunted" out of the closed cycle into the end products of differentiation, which accumulate. Soluble glycogen decreases with a consequent fall in the rate of synthesis and in the concentration of G1P and UDPG. Thus, an analysis of our initial model indicates that a change in the flux of G1P is the primary critical variable in affecting the rate of UDPG synthesis (Fig. 14E); a change in the concentration of UDPG pyrophosphorylase is not a primary critical variable (Fig. 14C). Secondary critical variables are those factors responsible for the increase in flux of G1P—e.g., changes in the value of $K2$ (i.e., the activity of glycogen phosphorylase), which is in a critical kinetic position (see below).

Table 10

THE RELATIONSHIP BETWEEN THE $[S]/K_m$ RATIO AND
THE EFFECT OF INCREASING THE V_v OF UDPG PYROPHOSPHORYLASE TEN-FOLD [29]

S/K_m	V_{UDPG_0}[a]	$V_{UDPG_{630}}$	$\dfrac{V_{630}}{V_0}$	$[G1P]_0$	$[G1P]_{630}$	$\dfrac{[G1P]_{630}}{[G1P]_0}$	$[UDPG]_0$	$[UDPG]_{630}$	$\dfrac{[UDPG]_{630}}{[UDPG]_0}$
0.001	2.8×10^{-3}	2.8×10^{-3}	1.0	1×10^{-5}	1.00×10^{-6}	0.1	5.6×10^{-5}	5.6×10^{-5}	1.0
0.01	2.8×10^{-2}	2.8×10^{-2}	1.0	1×10^{-5}	0.99×10^{-6}	0.099	5.6×10^{-5}	5.6×10^{-5}	1.0
0.1	2.6×10^{-1}	2.6×10^{-1}	1.0	1×10^{-5}	0.92×10^{-6}	0.092	5.6×10^{-5}	5.6×10^{-5}	1.0
1.0	1.4	1.4	1.0	1×10^{-5}	0.53×10^{-6}	0.053	5.6×10^{-5}	5.6×10^{-5}	1.0
5.0	2.4	2.4	1.0	1×10^{-5}	0.18×10^{-6}	0.018	5.6×10^{-5}	5.6×10^{-5}	1.0

[a] V_{UDPG} *is expressed as* $\mu moles/min/ml$ *packed cell volume.*

One of the interesting conclusions arising from this kinetic analysis is that changes over a wide range in the concentration of UDPG pyrophosphorylase could not affect the rate of UDPG synthesis: a reaction can go no faster than the rate at which substrate is supplied, regardless of the amount of enzyme present. Alterations in the activity or amount of this enzyme affected only the concentration of G1P. We wished to know if this result depended upon the fact that the $[S]/K_m$ ratio for this reaction was very low (0.01) [29]. In other words, is the $[S]/K_m$ ratio critical with respect to the effect on the system of increasing the V_v of UDPG pyrophosphorylase over the period from aggregation to culmination? Only the first 630 min were examined, so $K3 = 0$, and enzyme was increased tenfold rather than the threefold used previously (see Fig. 14C). Since cellular substrate levels were known and fixed, the $[S]/K_m$ ratio was varied by changing K_m. The rate constants, $K1$ and $K2$, were adjusted to produce a steady-state situation under the initial conditions. Table 10 shows the effect of increasing V_v tenfold on the concentrations of G1P and UDPG and on the rate of UDPG synthesis.

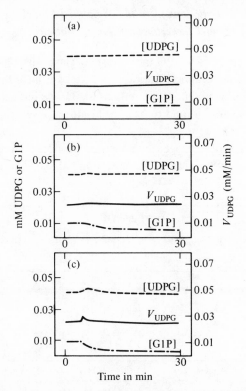

Figure 16. A computer simulation of the effect of increasing the V_v of *UDPG* pyrophosphorylase 10 fold over 630 (*A*), 100 (*B*) and 10(*C*) min., respectively. The S/K_m ratio was 0.01. In each case, only the first 30 min of computer output is shown [29].

The sole effect at all ratios examined was a compensatory reduction in the concentration of G1P required to establish a new steady-state. An additional examination of the time period over which a transient increase in the rate of UDPG synthesis would occur indicated that disturbance in the system is slight even if the tenfold increase takes place over 10 min and quickly disappears in about 5 to 10 min [29]. This is illustrated in Figure 16. Apparently, a change in the level of this enzyme is not a primary critical variable in affecting the rate of UDPG synthesis, even over a wide range of $[S]/K_m$ ratios. Studies such as the theoretical one just described bear upon the possible effect of compartmentalization of a $[S]/K_m$ ratio and hence upon the rate of a reaction *in vivo*. For reactions in a kinetic position such as UDPG synthesis, flux will in all probability be determined by substrate availability, regardless of the existence of compartmentalization [32].

Predictions from the Simplified Model and Experimental Substantiations

1. The activity of UDPG pyrophosphorylase must be inhibited *in vivo* if its effective concentration increases during differentiation; otherwise, G1P would not accumulate [30, 35]. As indicated in Table 11, UDPG has been found to be an inhibitor at levels present in the cell [14].

2. The rate of glycogen synthesis is proportional to the concentration of UDPG during differentiation [30, 35]. This assumption

Table 11

KINETIC CONSTANTS FOR UDPG PYROPHOSPHORYLASE

Kinetic constant	Value
K_{UTP}	1.1×10^{-4}M
$K_{i,UTP}$	1.1×10^{-4}M
K_{G1P}	3.2×10^{-4}M
K_{PPi}	3.0×10^{-5}M
K_{UDPG}	4.6×10^{-5}M
$K_{i,UDPG}$	4.6×10^{-5}M

Kinetic constants were determined at pH 8.0, 25°C at optimum magnesium concentrations (Mg^{+2} concentrations = twice concentration of either UTP or UDPG. $V_f/V_r = 0.50$ [14].

has been substantiated indirectly by the observation that exogenous glucose can briefly perturb the system, presumably by enhancing cellular levels of G1P and UDPG, resulting in a transient increase in the size of the glycogen pool [23]. A similar assumption (that the rate of G1P production is proportional to the concentration of soluble glycogen) has recently been challenged in the literature due to a misunderstanding of the kinetic model [2]. An observation that myxamoebae with higher glycogen levels show no increase in cellulose content at the end of differentiation would not bear upon the above assumption. The initial value (at aggregation) of V_v for glycogen phosphorylase (and amylase) is adjusted so that the rate of the reactions in the glycogen cycle are balanced—i.e., this calculated value would be lower if glycogen levels were higher. Other factors contributing to the amount of cell wall material formed are, for example, the amount of glycogen remaining at the end of differentiation, the value given the rate constant (K_v) for the end-product synthetase, and the relative competition of the trehalose synthetic pathway for common precursor materials (see expanded model below).

3. The rate of glycogen synthesis and degradation should equal V_{UDPG} at aggregation and should also increase about threefold between aggregation and culmination [30, 35]. This prediction concerning glycogen turnover was examined by two different experimental procedures [17]. First, cells were exposed to [^{14}C]-glucose over a period of time during which the specific radioactivity of intracellular UDPG was relatively constant and radioactivity of the glycogen pool was increasing in a linear fashion (Fig. 17). The rate of glycogen synthesis was calculated from the increase with time in μmoles of [^{14}C]-glucose incorporated, based on the specific radioactivity of UDPG. In the second procedure, which measured both glycogen synthesis and degradation, the glycogen pool was labelled early in differentiation by exposing cells to [^{14}C]-glucose. After washing the cells free of radioactivity, they were allowed to differentiate. Under these conditions glycogen specific radioactivity stayed constant for hours at any stage of differentiation even though protein specific radioactivity was much lower (Fig. 18). This observation supported other data (see Chap. 2) indicating little gluconeogenesis, which would tend to lower glycogen specific radioactivity with time. This result is also consistent with the existence of a closed "glycogen cycle" in

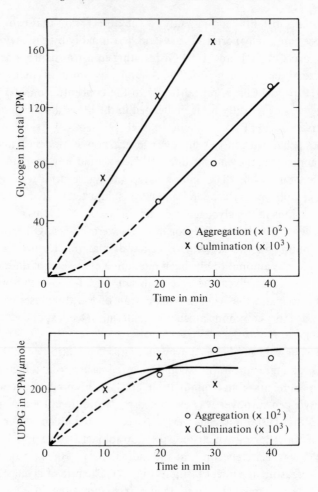

Figure 17. Glycogen synthesis at aggregation and culmination. See text for details [17].

which glucose units are cycled as indicated in the kinetic model (Fig. 13). However, when cells were "chased" at either aggregation or culmination with unlabeled glycose, the specific radioactivity of glycogen decreased (see Fig. 18). Knowing the concentration of intracellular glycogen at each stage of differentiation and the time required for glycogen specific radioactivity to reach one-half its initial value, it is possible to estimate the turnover rate in μmoles/glucose/min/ml. The rates of glycogen synthesis and turnover are compared to the rate of UDPG synthesis in Table 12; the differences in the three values indicated at each stage are not significant.

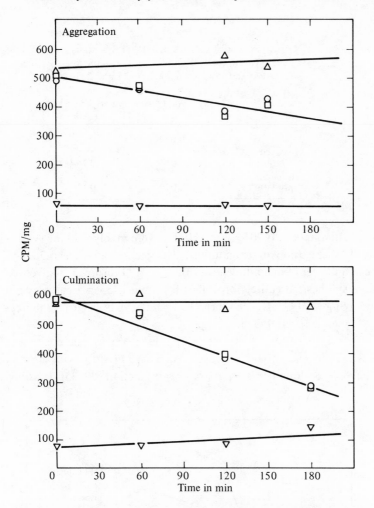

Figure 18. Relative radioactivities of glycogen and protein at aggregation and culmination: ☐, ⊙, glycogen, [¹²C]-glucose chase; △, glycogen, no [¹²C] glucose chase; ▽ protein, no [¹²C]-glucose chase. See text for details [17].

Support for the model was thus obtained by substantiating predicted flux values for glycogen synthesis and degradation. If UDPG is the only direct precursor of glycogen *in vivo*, then its rate of synthesis must be equal to or greater than the rate of glycogen synthesis. The fact that the values are similar is evidence in support of the assumption that UDPG is the real precursor of glycogen in the intact cell and that the synthesis of this polysaccharide is the major fate of UDPG

Table 12

A Comparison of the Rates of UDPG Synthesis,
Glycogen Synthesis, and Glycogen Degradation.
(For original data, see references 17 and 20.)

| | Flux ($\mu mole/min/ml\ p.c.v.$) | |
Reaction	Agg.	Culm.
V_{UDPG}	0.040	0.12
Glycogen synthesis	0.034	0.16
Glycogen degradation	0.046	0.13

---

until culmination. If 90 % of the UDPG were involved in the synthesis of other carbohydrate materials, glycogen degradation could have been an order of magnitude lower than UDPG turnover; if glycogen were synthesized entirely or in part from glucose via a pathway not involving UDPG, glycogen turnover could have been correspondingly higher than UDPG turnover.

4. Glycogen phosphorylase ($K2$) holds a key kinetic position in this system and should increase in activity; *in vitro* evidence may therefore be found in support of this prediction [30, 35]. Possibly,

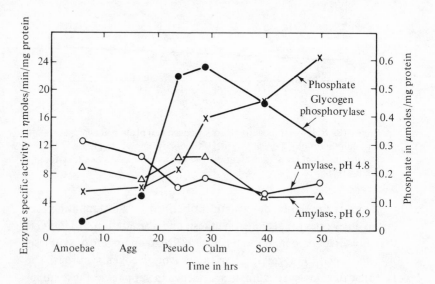

Figure 19. Variations of amylases, phosphorylases, and phosphate during development of *D. discoideum* [17].

changes in the activity of $K2$ could be attributed to an increase in the concentration of glycogen phosphorylase or its substrate, Pi. It was found that the enzyme did change in specific activity and that the concentration of Pi at aggregation [11] is close to the Km value of the enzyme for Pi [15]. Thus, a tenfold increase in Pi concentration could significantly alter the rate of glycogen degradation. Considering both phosphorylase and amylase activity, the increase in total glycogen-degrading capacity was more than sufficient to account for the threefold increase in the rate *in vivo* of glycogen degradation (Table 12 and Fig. 19) [15].

Expansion of the Model

The substantiation of a number of assumptions and predictions that were inherent in, or simulated by, the simplified model appeared to justify its expansion [32]. Based upon new experimental findings and upon information already available in the literature [3, 6, 7, 9, 14, 15, 22, 31] the model was refined to include the reactions indicated in Fig. 20. The accumulation

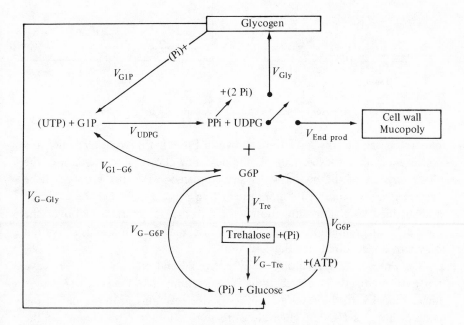

Figure 20. An expanded kinetic model of carbohydrate metabolism in *D. discoideum* [32].

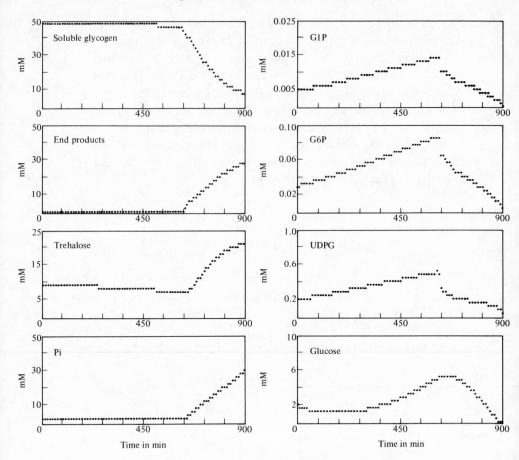

Figure 21. Computer output based on the expanded kinetic model. See text.

patterns based on this model are shown in Fig. 21; this computer output is consistent with the experimental findings. For clarity of presentation, the characteristics of the model which are consistent with all the data will be presented first. Some of its more interesting aspects may then be explored and justified. Activation functions ($V_v(t)$) and reaction rates (V) over the course of differentiation are given for those reactions in which activation functions are required (see Figs. 22 and 23).

V_{Gly} and $V_{End\ prod}$, resulting in glycogen and end-product synthesis, have not been modified. These reactions include simple rate constants (K_v) in a linear model with the rate of their respective reactions proportional to the concentration of UDPG. The rates of these reactions and the changes in the values of the rate constants as a function of time are given in Fig. 22. Treha-

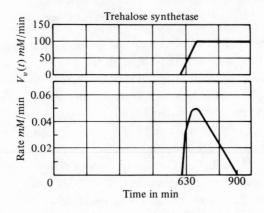

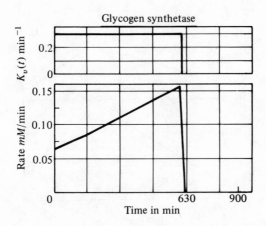

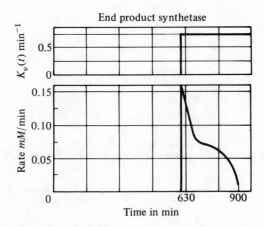

Figure 22. The rate of trehalose, glycogen and end product synthesis during differentiation, and the activation functions for the enzymes involved.

lose is now separated from the other end products and synthesized from UDPG and G6P by a separate pathway [26, 34].

V_{UDPG} catalyzed by UDPG pyrophosphorylase, is no longer described by a bimolecular initial velocity expression. The detailed reaction mechanism has been determined as ordered bi bi [13, 14], and the kinetic constants and cellular levels of substrates and end-product inhibitor are known (see Table 11). The reaction rate and enzyme activation function for this reaction are given in Fig. 23. A further characterization of this reaction may be seen in Figure 24, indicating the normalized rate of the reaction as a function of UDPG and G1P concentration. A linear response to increasing levels of

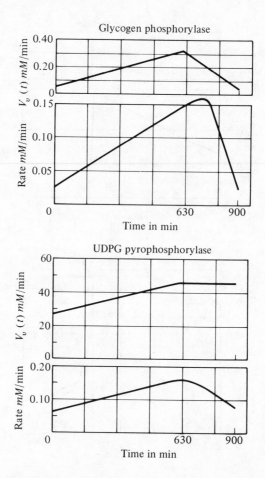

Figure 23. The rate of glycogen and UDPG synthesis during differentiation, and the activation functions for the enzymes involved.

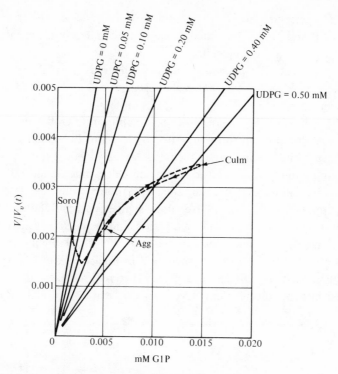

Figure 24. The normalized reaction rate for UDPG synthesis as a function of G1P and UDPG concentration. See text for further explanation.

G1P is seen at all levels of UDPG. The dashed line indicates the "operating track" over the course of differentiation by giving cellular levels of UDPG and G1P [32].

V_{GIP} represents G1P formation from glycogen and Pi, catalyzed by glycogen phosphorylase. The relevant substrate concentrations and K_m values are known [15]; this reaction is tentatively described using a bimolecular initial velocity expression, until data are available for the more accurate kinetic expression. At zero time (aggregation) this reaction accounts for one-half of the rate of glycogen degradation. As indicated in Fig. 19, the activity of glycogen phosphorylase increases sixfold between aggregation and culmination, then approaches 0 in older sorocarps. The reaction rate and activation function are given in Fig. 23. The "operating track" and reaction rate as a function of cellular substrate levels are indicated in Fig. 25.

V_{G-Gly} represents the rate of glucose formation from glycogen, catalyzed by amylase [15]; the reaction is represented by a linear rate equation, and its rate is constant throughout differentiation. At zero time this pathway is responsible for one-half of the rate of glycogen degradation.

$K_{G1\text{-}G6}$ represents phosphoglucomutase, which catalyzes the interconversion of G1P and G6P. This reaction has been handled in a special manner and is discussed in the Appendix.

V_{G6P} represents the phosphorylation of glucose by ATP, catalyzed by a specific glucokinase which has been characterized and may be described by an ordered bi bi initial velocity expression [3].

$V_{G\text{-}G6P}$ represents the rate of glucose formation from G6P catalyzed by an acid phosphatase known to be inhibited by levels of Pi present in the cells [11]. The inhibition constant is included in the kinetic expression, which is ordered uni bi [9]. It is assumed that glucose does not significantly inhibit this reaction, and terms containing this expression are omitted.

V_{Tre} represents the rate of trehalose synthesis from UDPG and G6P and involves two enzymes; T6P synthetase and a phosphatase converting T6P to trehalose. T6P does not accumulate [26]. Thus, T6P appears to be a small pool turning over rapidly (such as B in Figure 31A of the Appendix) and is suppressed. These two reactions are therefore represented as one and described by a bimolecular initial velocity expression. The activation func-

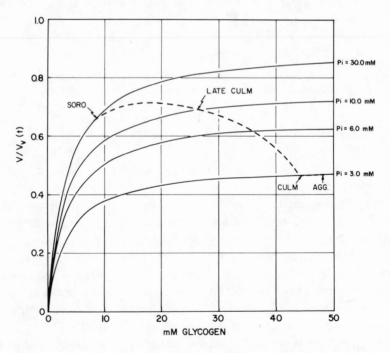

Figure 25. The normalized reaction rate for glycogen phosphorylase as a function of Pi and glycogen concentration. See text for further explanation.

tion and reaction rate are indicated in Fig. 22. Preliminary data indicate that G6P levels are well below saturation, and UDPG levels are approaching saturation during differentiation.

$V_{G\text{-}Tre}$ represents the rate of glucose formation from trehalose, catalyzed by trehalase; this enzyme has been characterized and is described by a uni bi initial velocity expression [6].

Independent metabolites, indicated by parentheses in the model (Fig. 20), act as substrates or products in the model, but are primarily controlled by reactions not explicitly included. Their concentrations are simply described as functions of time according to observed accumulation patterns (see Appendix).

UDPG pyrophosphorylase. We may now discuss a few aspects of the model in some detail. One prediction of the simplified model that was substantiated was that UDPG might inhibit the reaction. Having incorporated the accurate enzyme kinetic expression with the inhibition by UDPG into the model, the effects of changing the enzyme level may be explored again. Expansion of the model also may have affected earlier conclusions. Previous studies had indicated that, in the absence of both UDPG inhibition and an increase in V_v, the concentration of G1P and UDPG, as well as V_{UDPG}, increased about threefold from 0 to 630 min. Table 13 summarizes the results of increasing the enzyme level (V_v) one to fourfold over 630 min under the new conditions. The parameters listed are the change in concentration of UDPG and G1P and the change in the rate of UDPG synthesis as a function of changes in V_v.

Again, the main effect of increasing the enzyme level is to lower the steady-state level of G1P rather than to affect the level of UDPG or the rate of the reaction. As a small pool turning over rapidly, G1P levels vary readily in response to the rest of the system. In order to maintain the experimentally observed threefold accumulation of G1P in the presence of the inhibition by

Table 13

THE EFFECT OF AN INCREASE IN V_v FOR UDPG PYROPHOSPHORYLASE OVER 630 Min.

Fold increase in $V_v(t)$	$[UDPG]_{630}/[UDPG]_0$	$[G1P]_{630}/[G1P]_0$	$[V_{UDPG}]_{630}/[V_{UDPG}]_0$
1	2.3	4.6	2.3
2	2.6	2.7	2.6
3	2.7	1.9	2.7
4	2.8	1.6	2.8

UDPG, the enzyme level must be increased 1.7 fold; thus, the enzyme is given this activation function, as can be seen in Fig. 23. The effect of changing the $[G1P]/K_{G1P}$ ratio on the parameters listed in Table 13 was also explored using the expanded model with essentially the same results as had been obtained previously (Table 10, Fig. 16) [32]. These comparative studies indicate that, depending upon the question being asked, the precision of the enzyme kinetic expression may not be critical.

An analysis of the first model, in which this reaction was expressed by a simple initial velocity expression, revealed that the primary critical variable was the change in flux of G1P. This conclusion is again reached after incorporating the accurate enzyme kinetic expression with the inhibition by UDPG into the model. Only insofar as the level of G1P affects the rate of its production (e.g., by inhibiting the activity of glycogen phosphorylase), will changes in the level of UDPG pyrophosphorylase become a secondary critical variable for this reaction, by controlling G1P levels. That enhanced levels of this enzyme will lower G1P levels is shown by the computer analysis summarized in Table 13. Although G1P appears to inhibit glycogen phosphorylase [23], the extent of this inhibition is not yet known.

Glycogen phosphorylase Another prediction of the simplified model was that the rate of glycogen degradation should increase about threefold between aggregation and culmination and that *in vitro* evidence might be found for an enhanced activity of glycogen phosphorylase. Although both predictions were substantiated, the enzyme increased in excess of that required (fivefold or more). Furthermore, the accumulation of Pi towards the end of differentiation would also result in an increase in reaction rate. Yet the total glycogen-degrading capacity can increase no more than threefold between aggregation and culmination. If the initial activity is high enough to achieve the observed rate of glycogen degradation at aggregation, an increase in excess of threefold would give too high a rate at culmination—i.e., a rate greater than that determined *in vivo*. Therefore, assuming that glycogen degradation is mediated *only* by glycogen phosphorylase, we conclude that the latter enzyme does not increase in activity *in vivo* as indicated *in vitro*— i.e., is inhibited or partially masked.

Alternatively, amylase may be active *in vivo* and in part responsible for glycogen degradation. This, in fact, is assumed to be the case and is incorporated into the model, as it is consistent with the data obtained *in vitro* for both enzymes. That is, considered together, the change in their specific activity can account for the required threefold increase in the rate of glycogen degradation *in vivo*. Table 14 summarizes selected computer output for cases in which amylase activity accounts for one-half of the rate of glycogen degradation initially, and glycogen phosphorylase increases in activity two, four, six, or eightfold between 0 and 630 min. For the case in which a sixfold increase was simulated, the rate of glycogen degradation due to both enzymes increased threefold. A sixfold activation function between 0 and 630 min is therefore used (Fig. 23). Model exploration also indicated that the *decrease* in glycogen phosphorylase activity observed *in vitro* during sorocarp construction also occurs *in vivo*. If it did not, glycogen would be depleted prematurely. This decrease in enzyme activity partially offsets the increase in reaction rate due to Pi accumulation.

Table 14

THE EFFECT OF CHANGES IN V_v OF GLYCOGEN PHOSPHORYLASE BETWEEN 0 AND 630 MIN.

Fold increase in V_v	$[V_{G1y}]_{630}$	$\dfrac{[V_{G1y}]_{630}}{[V_{G1y}]_0}$	$[V_{G1P} + V_{G-G1y}]_{630}$	$\dfrac{[V_{G1P} + V_{G-G1y}]_{630}}{[V_{G1P} + V_{G-G1y}]_0}$
2	0.08	1.26	0.08	1.55
4	0.12	1.94	0.12	2.50
6	0.16	2.49	0.17	3.41
8	0.19	3.00	0.22	4.35

Apparently changes in both enzyme activity and substrate concentration are primary critical variables in increasing and decreasing the rate of glycogen degradation. The mechanisms underlying these changes, such as the affect of 3'5'-AMP on glycogen phosphorylase, become secondary critical variables (see p. 40).

Trehalose accumulation. It appeared from the literature that another potentially fruitful point for expansion would be an analysis of the mechanism of trehalose accumulation; that is, trehalose could be removed from the end products of the first model and be treated as a separate pathway [33, 34]. Ceccarini and Filosa have provided information on the pattern of trehalose accumulation and on the behavior *in vitro* of the enzyme responsible for its degradation, trehalase (V_{G-Tre}) [6, 7]. Roth and Sussman have described the behavior *in vitro* of T6P synthetase (V_{Tre}) [22]. These data are summarized schematically in Fig. 26 (note that trehalose concentration is expressed as percent dry wt.; see Table 4, Chap. 2); at about 500 min (preculmination) the synthetic enzyme has reached its maximum and the degradative enzyme its minimum specific activity.

A rate of trehalose synthesis, V_{Tre}, was assumed such that the proper amount of trehalose would accumulate over the required period of time. Based on this assumed rate, the cellular substrate levels and K_m values, the calculated value for V_v was 20 μmole/min/ml. The other reactions in the

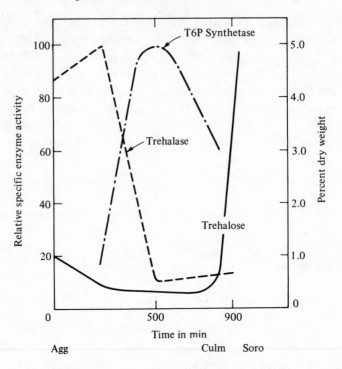

Figure 26. A schematic summary of the changes during
differentiation in the concentration of trehalose and in the
specific activities of trehalase and T6P synthetase [6, 7, 22].

"trehalose cycle" were accordingly balanced in order to maintain steady
state concentrations of the metabolites involved. In creating this pathway,
neither the accumulation patterns nor the flux values now established for the
"glycogen cycle" could be disturbed.

The purpose of expanding this aspect of the kinetic model was to explore,
simulating the steady-state conditions of the living cell, the effects of changing
the activity of the enzymes (V_v values) for trehalose synthesis and degradation
according to the *in vitro* data (Fig. 26). Therefore, initially, the synthetic
enzyme was increased tenfold between 0 and 500 min, and the degradative
enzyme decreased to 0 between 450 and 550 min.

"Turnover" model. Since some trehalose and the synthetic and degradative
enzymes were all present prior to the rapid accumulation of trehalose, the
first assumption in model construction was that trehalose would turn over
prior to 500 min. Output from the successful model—i.e., the one which
mimics known accumulation patterns and flux values—is shown in Fig.
21. Computer output indicated that this resulted in a rate of trehalose synthesis (V_{Tre}) of about 0.02 μmole/min/ml packed cells. The accumulation

patterns obtained were compatible with those observed experimentally; furthermore, flux values compatible with those of Table 12 were also found. This model incorporates the decrease in degradative enzyme activity indicated in Fig. 26, but holds the activity of the synthetic enzyme constant over the course of differentiation.

Now, let us examine the consequences of increasing V_v of V_{Tre} between 0 and 500 min, as observed *in vitro* in the presence and absence of changes in V_v of the degradative enzyme. In curve A of Fig. 27 the level of both en-

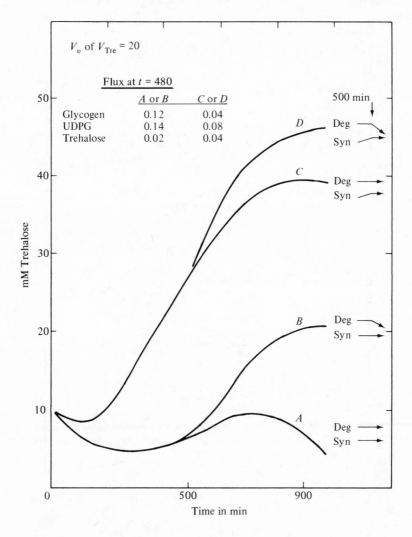

V_v of $V_{Tre} = 20$

Flux at $t = 480$

	A or B	C or D
Glycogen	0.12	0.04
UDPG	0.14	0.08
Trehalose	0.02	0.04

Figure 27. The consequences to trehalose accumulation of increasing the activity of the synthetic enzyme.

zymes was held constant and, therefore, trehalose did not accumulate to the proper extent. Modulation in its level resulted from the combined effect of the low level of degradative activity present, offset to some extent toward the middle of differentiation by the accumulation of the two precursors, G6P and UDPG. Curve B represents the output for the successful model for which the accumulation patterns of all components are shown in Fig. 21. Curve C resulted from an increase only in V_v of V_{Tre}: curve D resulted from an increase in V_v of V_{Tre} and a decrease in V_v of V_{G-Tre}. In cases C and D trehalose began to accumulate too early and did so in excessive amounts;

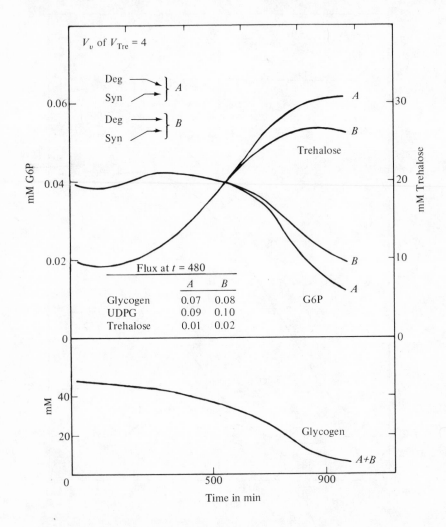

Figure 28. The consequences of increasing the activity of the synthetic enzyme, beginning with a lower initial value.

furthermore, the maximum rates of glycogen and UDPG synthesis, occurring at $t = 480$ min, were too low (compare the table in Fig. 24 with Table 12). Perhaps an unrealistic value for V_v of V_{Tre} was chosen; a lower initial value may have allowed a tenfold increase in activity without consequences incompatible with the data. In Fig. 28, the value V_v of V_{Tre} (and hence, necessarily, all trehalose cycle rate constants) was lowered by one-fifth to 4 μmole/min/ml. Under these new conditions, the synthetic enzyme was again increased, both in the presence (A) and absence (B) of changes in the degradative enzyme. The pattern and extent of trehalose accumulation was now acceptable—but three other results were not: (1) G6P did not accumulate; (2) too much glycogen was used; and (3) the maximum flux values for UDPG synthesis and especially glycogen synthesis fell short of the data determined *in vivo*. (All of these deficiencies also occurred in the output, giving curves C and D of Fig. 27.) Attempts to overcome some of these deficiencies (e.g., increasing the flux of G6P from G1P) either failed or resulted in other output incompatible with the data.

Exploration of these and related models forced us to the conclusion that only the decreased activity of the degradative enzyme (Fig. 26) can account for all the data when both enzymes are active, assuming that trehalose turns over prior to 500 min.

"No turnover" model. Models assuming no trehalose degradation prior to 500 min predicted that observed changes in the activity *in vitro* of the synthetic enzyme could not apply in *vivo*, even in the absence of degradative activity. Output from such a model is shown in Fig. 29. As a control, curves are shown with neither the synthetic nor degradative enzyme present (i.e., in the absence of the trehalose cycle). In the presence of synthetic activity, in order to achieve the proper amount of trehalose accumulation, the initial V_v of V_{Tre} was lower in the case in which the synthetic enzyme level was increased. Although these models improved the glycogen picture, G6P still did not accumulate properly, and trehalose accumulated too soon and in too linear a manner. Our analysis and conclusions assume that there is no noncarbohydrate source of G6P in this system—an assumption supported by the data of Cleland and Coe (see Chap. 2).

Assuming neither trehalose synthesis nor significant degradation prior to 500 min, a model was constructed which gave the proper accumulation patterns (i.e., those shown in Fig. 21 for a turnover model) by initiating synthetic activity *after* 500 min (e.g., at 630 min). In this model an extremely low and constant degradative activity was included from time zero (1/100 of that used previously) to account for the drop in trehalose level early in differentiation as shown in Fig. 26. The initiation of synthetic activity at 630 min coincides with and helps account for the decrease in steady-state level of

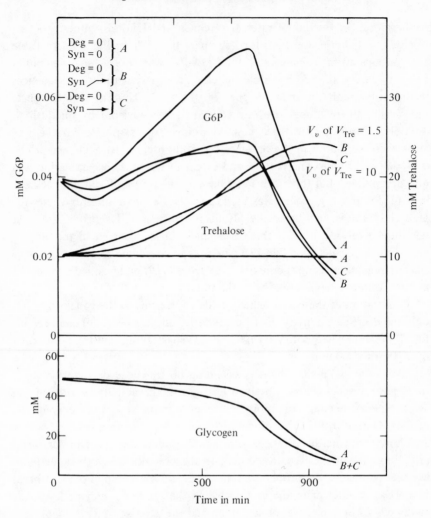

Figure 29. Output from models in which no degradative activity occurs.

the two precursors, UDPG and G6P. Either this model or the "turnover" model incorporating changes in the activity of the degradative enzyme satisfy the experimental data.

Predictions from the Expanded Model and Experimental Substantiations

1. *In vivo,* the effective concentration of UDPG pyrophosphorylase increases about twofold between aggregation and culmination.

A further increase must occur if cellular levels of PPi inhibit the reaction or if GIP inhibits the rate of its production (see p. 76).

2. Whether or not G1P is compartmentalized with respect to UDPG pyrophosphorylase, a change in the flux of G1P is the primary critical variable controlling the rate of the reaction.

3. If glycogen degradation is mediated only by glycogen phosphorylase, the latter enzyme does not increase in activity *in vivo* as indicated *in vitro*—e.g., it is inhibited (by G1P?) or masked.

4. If glycogen phosphorylase is not significantly inhibited or masked *in vivo*, then amylase is partially responsible for glycogen degradation.

5. The rate of production and utilization of inorganic phosphate has been predicted, based upon the reactions indicated in Fig. 20 and on calculations of the rates of protein and RNA turnover and of oxidative phosphorylation. It is of the order of 2 mM/min. This prediction is discussed in some detail elsewhere [32].

6. Observed changes in the activity *in vitro* of T6P synthetase cannot apply *in vivo;* that is, an increase in the rate of trehalose synthesis cannot occur prior to culmination. The actual rate of trehalose synthesis and the approximate time period over which a burst of about 100-fold in synthetic activity must occur were also predicted [33, 34].

Methods were developed for the isolation of trehalose, which was identified by chromatographic and enzymatic analyses as well as by its conversion to the octaacetate derivative and subsequent thin layer chromatography [26]. As in the experiments designed to determine the rate of glycogen synthesis *in vitro*, cells were exposed to [^{14}C]-glucose, and samples were analyzed every 10 min. The rate of trehalose synthesis was based upon the specific radioactivity of the precursors, UDPG and G6P, and upon the increase with time in radioactivity of the trehalose pool. Examination of the rate of trehalose synthesis *in vivo* (Fig. 30) revealed insignificant activity during differentiation until late in culmination at which time a burst of synthesis occurs, amounting to an increase of about 100-fold. Therefore, although *in vitro* data indicated that the synthetic enzyme, the degradative enzyme, and trehalose are all present prior to culmination (Fig. 26), trehalose turnover is negligible. No accumulation of T6P was observed. Had T6P turned over rapidly and not accumulated, this would have been reflected in the radioactivity of trehalose. Thus, a prediction of the expanded model was substantiated more than a year after it was made: Observed changes *in vitro* in the specific activity of the synthetic enzyme cannot and do not reflect activity *in vivo*.

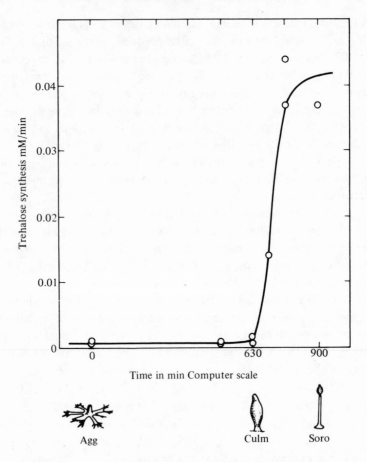

Figure 30. The rate of trehalose synthesis determined *in vivo*. See text
for details.

Exploration of our "no turnover" model [34] predicted that a burst of approxi-
mately 100-fold in synthetic activity must be initiated late in culmination, occur
over a relatively short period (about 2 hr), and be of the order of 0.03 mM/
min (see Fig. 22). During this burst in synthetic activity trehalase cannot be
active, or a greater rate of synthesis would be required to achieve the proper
level of trehalose accumulation. Thus, at no time during differentiation is
trehalase very active; glucose accumulation probably arises largely from
phosphatase action on the hexose phosphates as this enzyme and its substrates
peak when glucose does [9]. Even though amylase activity does not peak at
this time [15], glucose may also arise from this source.

   The accuracy of our predictions concerning both the pattern of synthetic
enzyme activity and the rate of trehalose synthesis validates the now numer-

ous assumptions on which the kinetic models were based; had these assumptions been unrealistic, the experimentally determined rates of trehalose synthesis could have deviated from those predicted by orders of magnitude. In turn, substantiation of the predictions strengthens the validity of the data obtained *in vivo*. The primary critical variable in the initiation of trehalose synthesis late in culmination is an *"in vivo* unmasking"—i.e., although the substrates are present, the enzyme is inactive *in vivo*. Possible mechanisms include the activation or synthesis of T6P phosphatase or the sudden availability of substrates to enzyme (i.e., decompartmentalization). An analysis of this mechanism may be greatly facilitated through an understanding of

Table 15

CRITICAL VARIABLES FOR REACTIONS ESSENTIAL
TO DIFFERENTIATION IN *D. Discoideum*

Reaction	Primary	Secondary
Increased G1P formation (agg to culm)	(1) enhanced levels of active phosphorylase (2) increased activity of phosphorylase (3) increased levels of Pi	(1) conversion from inactive form by 3'5'AMP (?) (2) inhibition by G1P overcome by 5'AMP (3) decreased rate of oxidative phosphorylation (?)
Increased UDPG synthesis (agg to culm)	(1) increased rate of G1P availability	(1) primary critical variables (a, b, c) for G1P formation (2) increased levels of UDPG pyrophosphorylase (?)
Increased glycogen synthesis (agg to culm)	(1) increased UDPG and G6P availability	(1) increased G1P availability
Decreased glycogen synthesis (culm to soro)	(1) decreased UDPG and G6P availability (2) inhibition by Pi	(1) decreased G1P availability (2) decreased rate of oxidative phosphorylation (?)
Accumulation of cellulose-glycogen cell-wall complex (culm to soro)	(1) insolubilization of precursors (2) shift of glycogen synthetase from soluble to insoluble primer	(1) accumulation and increase in size of precursors (2) drop in level and rate of synthesis of soluble glycogen
Trehalose synthesis (culm to soro)	(1) *"in vivo* unmasking" of T6P synthetase	(1) rate of UDPG and G6P availability at 630 min (2) activity of *"in vitro* unmasked" T6P synthetase at 630 min

the "*in vitro* unmasking" mechanism (see Chap. 2). Secondary critical variables are: (1) the levels of the two precursors, UDPG and G6P, and their turnover rates, which are known to be adequate for and consistent with the amount and rate of trehalose accumulation; and (2) the activity of the "*in vitro* unmasked" enzyme discussed on p. 43.

By way of summary, our current understanding of control mechanisms in this system is indicated in Table 15 which suggests primary and secondary critical variables for some of the reactions thus far analyzed (Fig. 20).

Generalizations and Conclusions

Returning now to the central dogma of developmental biology discussed in the first chapter, how do our conclusions reflect upon the basic assumptions of this dogma? In the case of UDPG pyrophosphorylase, although the enzyme activity and reaction rate increase together during differentiation, the increased enzyme level is not a primary critical variable—i.e., is not directly responsible for the enhanced reaction rate. In the case of T6P synthetase, an increased enzyme specific activity is not accompanied by an increased rate of trehalose synthesis. These examples along with a number of others in this system [30] illustrate that changes in enzyme activity are *by no means* always primary critical variables. On the other hand, the change in activity of glycogen phosphorylase during differentiation *does* appear to be a primary critical variable. *In vitro* analyses alone appear to give us very little information regarding control mechanisms *in vivo*. Many comparisons indicate little correlation between the activity of an enzyme measured *in vitro* or activity changes during differentiation with estimated *in vivo* enzyme activity. Computer simulation can serve as a valuable tool with which to assess the pattern of enzyme activity *in vivo* and the consequences of such activity to the metabolic system as a whole. To have significance to the mechanism of differentiation, *in vitro* data must be accompanied by *in vivo* analyses and by a knowledge of the kinetic position of the enzyme in question. An increase in enzyme level may be incidental to and *result* from enhanced metabolic activity. Such an effect could be brought about by the increased flux of precursors and products capable of inducing or stabilizing the enzyme. A possible beneficial consequence of enzyme accumulation would be to insure rapid and thorough utilization of limited precursor materials during some future period of the differentiation process.

Most of the enzymes essential to energy metabolism and end-product accumulation during differentiation in *D. discoideum* are also present during the growth phase of the life cycle and are not critical variables over our chosen time period of analysis. These include enzymes common to most living cells: those of the TCA cycle, amino acid metabolism, the Embden–Meyerhof pathway, oxidative phosphorylation, and the enzymes responsible for the maintenance and turnover of the bulk of macromolecules in the cell. Some 20 enzymes of this type have been studied in *D. discoideum;* they are among those listed in Tables 7 and 8 in Chap. 2. Of those known to change significantly in level or location (Table 8 in Chap. 2), at least five appear not to be primary critical variables—i.e., the specific activity changes are not reflected by or critical to parallel changes in reaction rates *in vivo* (soluble glycogen synthetase, cellulase, T6P synthetase, UDPG pyrophosphorylase, and trehalase). If cellulase, trehalase, and T6P synthetase accumulate in an inactive form, changes in their level undoubtedly represent secondary critical variables. Changes in the activity of glycogen phosphorylase and the location of cell-wall glycogen synthetase do appear to be primary critical variables as does the release of trehalase activity during germination (as opposed to its activity changes during differentiation). It is noteworthy that the substrates for both glycogen phosphorylase and trehalase are present at maximum concentrations prior to the increase in enzyme activity so that enhanced precursor flux cannot be involved in the mechanism of increased activity. It is also of interest that both of these enzymes occupy key positions in the flux of metabolites into anabolic and catabolic pathways, respectively. During the actual time of end-product accumulation, the rate of precursor flux must always be a critical variable, as a stoichiometric relationship exists between precursor and product.

The exploration of our simple kinetic models has suggested a few tentative generalizations concerning the process of model construction and control mechanisms in the living cell. In simulating metabolic pathways under steady-state conditions, specific kinetic expressions were used in order to estimate the enzyme activity (V_v) from the observed flux *in vivo* (V), the substrate concentrations, K_m and K_i values, and from accumulation patterns. At this stage in our analysis, the initial, absolute substrate levels and K_m values are not particularly critical, in that variations in these parameters will be compensated for by the value of V_v, which is calculated. However, *changes* in substrate levels or K_m values as a function of time are very critical to an analysis of the control mechanisms involved. For example, when *in vitro* data indicate that an enzyme specific activity changes with the same pattern as its substrate (e.g., UDPG pyrophosphorylase, cellulase, T6P synthetase), then this increased enzyme activity probably does not occur *in vivo*, for this would be opposed to substrate accumulation. The size of a small rapid turnover pool immediately adjusts to small changes in the flux of reactions forming and using it. Enhanced activity of an enzyme using such a pool is quickly reflected in a decreased steady-state level of the pool.

The most important kinds of information in model construction appear to be a knowledge of (1) the detailed reaction sequences involved and the kinetic positions of the reactions, (2) key *in vivo* flux values; and (3) pooling patterns of metabolites. Information on metabolite profiles during differentiation are available for a few other systems potentially simple enough for analysis; intriguing examples are amoebic encystment [19], ascospore activation in Neurospora [27], Basidiospore germination in *Schizophyllum commune* [1], and differentiation in *physarum polycephalum* [18].

Conceptually, the very process of constructing and exploring kinetic models has given us a new dimension of understanding; such models represent a technique by which, theoretically, the dynamics of metabolism in a living cell may be examined *in toto*. Steady-state conditions may be simulated, and *in vitro* data may be critically evaluated for their relevance to the intact cell. It frequently is not possible to obtain the rate of a reaction *in vivo*, depending upon variables such as relative pool sizes, turnover rates, isolation procedures, and pertinent biochemical techniques. In the event that such a rate cannot be determined directly, a kinetic model may represent the only means by which it may be estimated. The multiplicity of control mechanisms and interdependent pathways in any cell are of such a complexity that they cannot be simultaneously comprehended by the human mind. We can, however, gather the data to be simulated and integrated by a kinetic model. This analytical approach acknowledges more variables essential to differentiation than have ever before been considered. The heuristic value of the primitive models used thus far has surpassed our expectations and has provided a rational basis on which to proceed experimentally.

In conclusion, we have examined the effect of changes in the activity of various enzymes and the flux of endogenous metabolites on differentiation and have found that the latter may play a surprisingly prominent role among the biochemical events underlying differentiation. However, it would be as misleading to define differentiation in terms of changes in metabolic flux as to define it in terms of changing patterns of protein synthesis. Both definitions represent viewpoints which are vastly oversimplified and hence detrimental to an unprejudiced analysis of the problem. With respect to the mechanism of differentiation, the significance of changes in enzyme levels or metabolite flux cannot be assessed in the absence of a detailed knowledge of all processes coupled to it, including its kinetic position in the system. A system of coupled interacting biological components has properties beyond those apparent from the analysis of these components in isolation. A hierarchical scheme of control, based on any single cellular component, is inadequate to account for biochemical behavior that necessarily owes its very existence to complex relationships among many components. The problem is to understand these relationships without isolating the steps of which they are composed.

Appendix

Programming features

METASIM is written in the programming language PL1 for the 1BM-360 series computers. It operates in 164 K bytes of core storage and utilizes no peripheral storage devices. The program is expressed in standard biochemical nomenclature both in the internal programming and in the input-output system. The program has an extensive data checking facility: All input data are checked for obvious errors and this checking continues regardless of the number of errors. The program provides a formatted printout of each reaction comprising the metabolic system and also allows ample descriptive information to be entered so that the output of each stimulation is self-contained and fully documented. The principal output is a tabular listing of reactant concentrations and reaction rates as a function of time. The program also provides printer plots of the reactant concentrations. Multiple simulations in one computer run may be performed by entering a base set of data for the first simulation, and then in each succeeding simulation entering only the parameters to be changed. The program also contained an automatic calculation time limit which protects the user against runaway simulations and has a restart procedure for simulations which were not finished within the time limit. A user's manual is available.

Metabolic features

METASIM will accommodate metabolic systems that contain as many as 25 enzymes and 25 reactants. Reactants and enzymes are separately described. Part of each enzyme description is a list of substrates, products, and effectors for the enzyme. The program automatically links the enzyme kinetic variables with the reactant variables to establish the correct differential equations. The program contains built-in enzyme kinetic equations with provision for adding more if required. These include a linear equation (rate = constant $\times$ reactant concentration), initial velocity equations, and many rate equations including inhibition constants in the notation of W. W. Cleland [8]. The user must pick the enzyme kinetic expression for each reaction and provide the maximum velocities (V_v), Michaelis binding constants, and inhibition constants.

To be effective a metabolic simulator must allow the user to segment a metabolic system in order to study a portion of it. The program has several features to aid in this process. External (i.e., arbitrarily added and not generated by the model) fluxes of metabolites may be specified as time functions. This allows the user to account for the fact that metabolites may be consumed or generated by reactions outside the portion of metabolism being studied. The concentrations of certain metabolites in the system may be primarily determined by reactions outside the metabolic segment being studied, yet these metabolites may be substrates for reactions in the system. Such metabolite concentrations can be specificed as time functions; in this case, they simply act as boundary conditions for the system being studied.

METASIM used a first order integration method for solving the differential equations. Higher order integration methods are of little value in performing metabolic simulations of this type. The integration step size is rigorously determined by the fastest metabolite pool turnover time in the system. Greater simulation efficiency can be obtained by suppressing metabolites with rapid turnover times. If the rapid turnover pool is one of a chain of metabolites in a metabolic pathway, as in pool B of Fig. 31A, then it may be suppressed by omitting both the pool and the enzyme that converts it to the next metabolite in the chain. The case of phospoglucomutase may be represented by Fig. 31B; G1P (A) and G6P (B) represent two rapid turnover pools that are connected by a reversible reaction carried out by enzyme e_1. If the e_1 reaction is much faster than reactions yielding or consumming A or B, then the ratio of the concentrations of A and B will be kept at the equilibrium value $K_{eq} = B/A$, and the net reaction rate of e_1 will be zero. There will still be rapid exchange between A and B, and the flux from A to B will be equal to the flux from B to A. These large fluxes contribute to the rapid turnover times of the A and B pools. This rapid turnover can be suppressed by lumping the two pools together into one pool AB. The AB pool is then integrated and the concentrations of the A and B pools are calculated by

$$A = (1/(K_{eq} + 1))AB$$
$$B = (K_{eq}/(K_{eq} + 1))AB$$

for purposes of evaluating the rate equations which require A and B. In some simulations we have used this approximation to lump G1P and G6P.

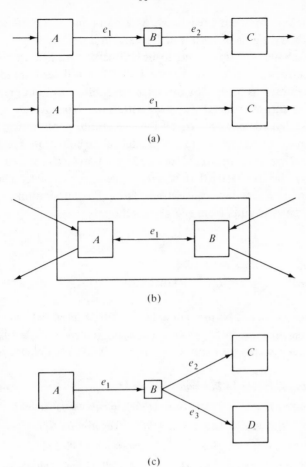

(a)

(b)

(c)

Figure 31.

Experimental data indicates that G6P $\cong 6.0 \cdot$ G1P throughout the course of differentiation [20]. If these two pools are integrated separately, it is necessary to use an average integration step size of 0.09 min in order to maintain a stable integration. Lumping the G1P and G6P pools raises the average step size to 1.4 min, resulting in a great increase in efficiency.

If the rapid turnover pool is at a branch point in the metabolic pathways, as for pool B in Fig. 31C, then the pool cannot be simply suppressed because its concentration determines the division of the flow of material at that branch point. The division will, in general, be nonstoichiometric and will depend upon the variation in the pool concentration and on the rate equations for the two competing reactions. A case in point is the competition for UDPG in the synthesis of both trehalose and end products.

The metabolic equations tend to be self-correcting against roundoff or truncation errors. Specifying too large a step size will cause random errors in the concentrations of the fast turnover metabolites without greatly affecting the slow turnover metabolite concentrations. The overall response of metabolic systems is constrained by the metabolite pools which have slow turnover times. For example, in our model of carbohydrate metabolism in this differentiating system, trehalose and other polysaccharide end products cannot be synthesized faster than substrate becomes available. The enzymes involved are nowhere near saturation. Thus, the rate-limiting step is the removal of glucose units from the glycogen pool.

REFERENCES

1. AITKEN, W. B. and NIEDERPRUEM, D. J. (1970). *J. Bact.* **104**, 981.

2. ASHWORTH, J. M. (1971). *Control Mechanisms of Growth and Differentiation* (Symposia of the Society for Experimental Biology XXV, 27). Cambridge: University Press.

3. BAUMANN, P. (1969). *Biochemistry* **8**, 5011.

4. BAUMANN, P. and WRIGHT, B. E. (1968). *Biochemistry* **7**, 3653.

5. BAUMANN, P. and WRIGHT, B. E. (1969). *Biochemistry* **8**, 1655.

6. CECCARINI, C. (1967). *Biochim. et Biophys. Acta* **148**, 114.

7. CECCARINI, C. and FILOSA, M. (1965). *J. Cell. Comp. Physiol.* **66**, 135.

8. CLELAND, W. W. (1963). *Biochim. et Biophys. Acta* **67**, 104.

9. GEZELIUS, K. (1966). *Physiol. Plant.* **19**, 946.

10. GEZELIUS, K. (1968). *Physiol. Plant.* **21**, 35.

11. GEZELIUS, K. and WRIGHT, B. E. (1965). *J. Gen. Microbiology* **38**, 309.

12. GRAY, J. S. (1963). *Science* **140**, 464.

13. GUSTAFSON, G. and WRIGHT, B. E. (1971). *Fed. Proc.* **30**, 1069.

14. GUSTAFSON, G. and WRIGHT, B. E. *CRC Critical Reviews in Microbiology* **1** issve 4 (A. I. Laskin and H. Leehevalier, eds.). Ohio: CRC Press. p. 453.

15. JONES, T. H. D. and WRIGHT, B. E. (1970). *J. Bact.* **104**, 754.

16. KASCER, H. (1963). In *Biological Organization at the Cellular and Super-cellular Level* (R. J. C. Harris, ed.). New York: Academic Press Inc.

17. MARSHALL, R., SARGENT, D., and WRIGHT, B. E. (1970). *Biochemistry* **9,** 3087.

18. McCORMICK, J. J., BLOMQUIST, J. C., and RUSCH, H. P. (1970). *J. Bact.* **104,** 1110.

19. NEFF, R. L. and NEFF, R. H. (1969). In *Dormancy and Survival XXIII S. E. B. Symposia.* New York: Academic Press Inc., p. 51.

20. PANNBACKER, R. G. (1967). *Biochemistry* **6,** 1283.

21. PARK, D., RUTHERFORD, C., and WRIGHT, B. E. (submitted for publication).

22. ROTH, R. and SUSSMAN, M. (1968). *J. Biol. Chem.* **243,** 5081.

23. ROSNESS, P. A., GUSTAFSON, G., and WRIGHT, B. E. (1971). *J. Bact.* **108,** 1329.

24. RUTHERFORD, C., KONG, W. Y., and WRIGHT, B. E. (submitted)

25. RUTHERFORD, C. and WRIGHT, B. E. (1971). *J. Bact.* **108,** 269.

26. SARGENT, D. and WRIGHT, B. E. (1971). *J. Biol. Chem.* **246,** 5340.

27. SUSSMAN, A. S. (1969). *In Dormancy and Survival* (Symposia of the Society for Experimental Biology XXIII). New York: Academic Press Inc., p. 99

28. WARD, C. and WRIGHT, B. E. (1965). *Biochemistry* **4,** 2021.

29. WRIGHT, B. E. (1970). *Behav. Sci.* **15,** 37.

30. WRIGHT, B. E. (1968). *J. Gen. Physiol. Supp.* 1, **72,** 145.

31. WRIGHT, B. E. and DAHLBERG, D. (1967). *Biochemistry* **6,** 2074.

32. WRIGHT, B. E., GUSTAFSON, G., and PARK, D. (1972). (submitted).

33. WRIGHT, B. E. and MARSHALL, R. (1969). *Biophys. Jour. Abs.* **9,** A-66.

34. WRIGHT, B. E. and MARSHALL, R. (1971). *J. Biol. Chem.* **246,** 5335.

35. WRIGHT, B. E., SIMON, W., and WALSH, B. T. (1968). *Proc. Nat. Acad. Sci. U. S.* **60,** 644.

4

Speculations on the Evolution
of Flux Control in Differentiation

"*Only many, many millions of years after the origin of life, when the complexity
of the organization of living bodies had increased greatly and when the problem
of the more or less accurate self-reproduction of this organization, therefore, could not be
solved satisfactorily by a single reaction constantly taking place within the system,
did there arise the necessity for creating a new mechanism which would guarantee the essential
conservatism of the living system. DNA, with its great metabolic inertia, was such
a mechanism. Thus, one may suppose that the appearance of DNA only
became necessary when the development of living bodies had already reached
a comparatively high level.*"

A. I. Oparin, 1961 [34]

The previous chapters have documented the importance to differentiation of changes in the flux of endogenous metabolites. We may now briefly consider certain characteristics of simple differentiating systems that are intimately related to endogenous substrate control and finally speculate on the successive stages through which these characteristics could have evolved.

Differentiation as an Isolated System

The conditions stimulating differentiation in primitive systems frequently are coupled with the isolation of these systems from environmental influences. Sporulation in bacteria, fruiting body formation in the cellular slime molds [39] and myxobacteria [12, 49], amoebic encystment, [33] zygote formation in *Chlamydomonas* [25, 40], and perithecia construction in fungi [3, 51], are in part initiated by some degree of nutritional deficiency. In these and other simple systems, availability of an energy source during the early stages of differentiation can reverse the process completely. At a more complex level, sexual differentiation can be inhibited by growth; indeed, as early as 1888 Maupas [30] observed that ciliates mate only when deprived of food. In plants, reproductive organs are normally formed only after growth has ceased; physiological isolation of plant tissue as the stimulus for differentiation has been emphasized by Steward and others (for a review see Torrey [47]).

Decreases in permeability may accompany certain stages of differentiation in the life cycle of a number of the simpler, more primitive systems, thus preventing the net uptake of exogenous nutrients in addition, perhaps, to the loss of precious and limiting endogenous energy sources [4, 53]. (Net uptake is not necessarily indicated by the incorporation of radioactive compounds into the differentiating cells: Reciprocal exchanges of endogenous with exogenous labeled materials can occur with no influence on the actual size of the endogenous pool [54].) In more complex, highly organized differentiating systems, the exclusion of external influences becomes more extreme—amphibian eggs are almost completely impermeable, as are the seeds of plants, and the eggs of reptiles and birds; insects and tadpoles undergoing metamorphosis are also relatively impervious to external influences [13, 24]. All of these systems are isolated and "closed" *in the sense* that they do not depend on exogenous nutrients, but start with a fixed level of specific endogenous reserves. The latter may exist within each cell (e.g., the cellular slime molds) or be separated from the differentiating organism (e.g., the yolk of eggs). These energy sources and precursor materials are, for the most part, used in an orderly and sequential manner and exhausted by the time differentiation is accomplished. In the chicken embryo a continuous decrease in the rate of cell division accompanies a progressive depletion of reserve materials. A general treatment of the dependence of differentiation on the metabolism of endogenous materials has been published [52]. In summary, the isolated system can vary from single, starving slime mold cells to the avian egg or germinating seed—yet in all cases there occurs an almost complete exclusion of influences from the external environment.

On the Apparent Competition between Growth and Differentiation in Primitive, Isolated Systems

In the microbial world an apparent incompatibility between growth and differentiation is strongly suggested: There are organisms, such as the cellular slime mold, in which the growth phase of the life cycle is completely separated from the differentiation phase. As mentioned above, this pattern is also seen in myxobacteria, fungi, and other microorganisms. With respect to plants there are conditions under which differentiation can occur independently of growth, which is usually inhibited in some way prior to the specialization of various tissues and reproductive organs [16]. This independence of cytodifferentiation from cell division is well illustrated in the case of γ-irradiated wheat. The treatment is so intense that extensive chromosome breakage occurs, completely preventing DNA synthesis and cell division. Nevertheless, such wheat grains can germinate normally at the expense of endosperm reserves and develop into typical seedlings. These "gamma plantlets" produce highly differentiated cells and tissues (e.g., trichoblasts, vascular elements, chlorenchymatous mesophyl and cortical parenchyma) from existing presumptive cell types present prior to irradiation. In addition to performing photosynthesis and net protein synthesis, they accumulate specific types of protein and cell wall materials not previously detected [19].

In many systems the metabolic capacities of a cell may be insufficient to handle the demands of rapid proliferation as well as the synthesis of specific products characteristic of the differentiated cell, particularly if a competition for common intermediates (e.g., ATP) occurs between biosynthetic pathways required to produce more cells and those required for the specialization of existing cells. The reciprocal relationship between cell division and differentiation induced by Ca^{++} in plant tuber slices suggests such a competition [2]. The extent of the incompatibility between the metabolism resulting in growth and that required for cytodifferentiation may depend upon the extent to which the system in question is limited by nutrient availability and upon the similarity of precursors for the products of the two types of metabolism. In contrast to certain microbial and plant systems, competition between these two kinds of metabolism has apparently been partially or totally overcome in the evolution of more complex and versatile organisms [1, 7, 41]. However, the metabolic behavior of these advanced forms may be better understood in light of their evolutionary history exemplified, perhaps, by differentiation of microbial systems.

In summary, a number of studies in more primitive differentiating systems tend to support the generalization that when such cells are no longer able to grow they frequently channel their metabolism into forming specific products resulting in cytodifferentiation. In general, this occurs when cell growth becomes limited by nutrient availability or when cells become mature, imbedded, or closed off from their environment, that is, when they become nutritionally and physically isolated. This observation is particularly applicable to simpler organisms, which will, of course, serve as our main examples in discussiong the evolution of differentiating systems. We shall now examine the consequences of physiological isolation to the metabolism of differentiation and then speculate on the evolution of such systems.

Metabolism in an Isolated System

What are the general metabolic characteristics of an isolated differentiating system? Each system starts with specific endogenous reserve materials on which it is dependent for all the catabolic and anabolic pathways essential to survival and to the unique transformations that occur. Under the conditions of this "programmed starvation," organisms capable of differentiation would be expected to have the ability to use their reserves both extensively and efficiently. Direct measurements of macromolecular reserves and respiratory quotients in such systems have been used to indicate the nature and the extent of utilization of these materials [52]. Thus, the activity of enzymes such as proteases, carbohydrases, and lipases is of primary importance to substrate control in differentiation. Hydrolytic enzymes may be introduced by the sperm at the time of fertilization [28] and are known to be activated or synthesized in the early stages of development in sea urchin eggs [5, 15, 27], seeds [9, 31], amphibia [6, 20], and sporulating bacteria [26]. Possible mechanisms for the induction or release from inhibition of such enzymes as a consequence of starvation have been discussed in Chap. 1; a general mechanism for the activation or synthesis of enzymes unique to differentiation may be the disappearance of specific inhibitors or repressors upon starvation. The appearance of hydrolytic enzymes in the early phases of differentiation may also have consequences such as the activation of other enzymes present in an inactive form [31], the release of "masked" mRNA in fertilized sea urchin eggs [32], or the selective removal of histones from chromosomal RNA during seed development [23, 44].

The availability of a specific amount of endogenous reserve material is a factor determining the period of time over which differentiation occurs and also in part controls the pattern of appearance in time of the intermediates and end products accumulating in the system. Moreover as shown in Chaps. 2 and 3, many of the changes associated with the ordered depletion of specific endogenous reserves may actually help to initiate and terminate the differentiation process.

With respect to metabolic control, the endogenous nature of differentiation in more primitive systems gives rise to a condition insuring the reproducibility and perpetuation of the system: multiple controls of metabolic pathways at the substrate level. Examples of this have been described earlier in the interactions that control cell wall synthesis in the cellular slime mold. Potentially hazardous variations in the levels of critical endogenous precursors can be met by compensatory mechanisms which depend upon the fact that substrates and effectors exist at limiting levels in the cell and that competing, interdependent reactions are involved [53]. For example, at low levels of UDPG, G6P exerts a greater stimulation of soluble glycogen synthesis than at high levels. Such a system has "damping" characteristics, providing an elaborate arsenal of alternatives with which to compensate for or take advantage of various adverse or favorable circumstances. In summary, metabolism in an isolated system creates at least three conditions favoring the stability of differentiation: (1) a reproducible, "programmed" utilization of a fixed amount of macromolecular reserves; (2) a set pattern of changes in the small molecular milieu of the cell, which in part initiates and terminates the activity of metabolic pathways essential to differentiation; and (3) a complex interaction of many limiting factors that tends to "buffer" the course of metabolism, such that the premature or delayed accumulation of an essential metabolite neither triggers nor prevents differentiation.

Speculations on the Evolution of the Isolated System

Are the characteristics of differentiating systems outlined above compatible with the selection of such a system in evolution? What might be the first steps in the transformation of a fragile cell, protected only by a membrane, into one capable of some simple form of differentiation, such as spore or cyst formation? One might first consider the possible selective advantages of such a transformation: survival over long periods under conditions of poor nutrition, high temperature, dessication, osmotic shock, etc. The maintenance of viability under such adverse circumstances could be better achieved by cells capable of independence from their environment. One of the first steps in evolution, therefore, might have been the selection of cells having high levels of endogenous reserves as well as the ability to use them efficiently and completely. Later in the evolutionary process, the acquisition of partial permeability barriers might protect the cell from toxic materials and prevent the loss of essential small molecules arising from the degradation of macromolecular reserves. Thus, a second major step in the evolution of differentiation could then be the selection of cells capable of making a protective cell wall or spore coat from endogenous materials under starvation conditions. A third step, necessary to the evolution of more complex forms, would confer upon such isolated systems the ability to grow as well as differentiate. If the evolution of differentiation is currently in progress, it should be possible to find in nature representatives of some of these early stages in the evolution of systems dependent upon endogenous substrate control

Starvation and survival

There is a vast literature documenting a correlation between the amount of a reserve material such as glycogen in a microbe with its ability to survive during subsequent starvation [10, 11, 17, 45, 50]. Of particular interest in the present context are studies on a strain of *Aerobacter aerogenes* capable of extended survival under starvation conditions [21]. Over a 4-hr period of starvation in phosphate buffer, this cell type was 100% viable as compared to a 37% loss for the parent strain. The varient was found to be smaller in size and lower in growth rate (1.9 compared to 2.5 divisions per hour). During starvation the varient metabolized a higher proportion of RNA and protein. Thus, the decrease in dry weight of this strain (19%) was greater than that of the wild type (13%). When a comparable loss in dry weight of the wild type was achieved by 19 hours of starvation, less than 1% of the cells survived. A metabolic change resulting in the activation of an enzyme-degrading ribosomes, for example, could be responsible for the behavior of this new strain. Proteases and ribonucleases frequently exist *in vivo* in an inactive state [8, 14, 20, 27] and can be activated on starvation [11, 18, 45]. An increased capacity of this nature could favor survival of the varient under starvation conditions, yet inhibit growth under optimal conditions by interfering with the rate of accumulation of cellular RNA or protein.

The conflict in the selection of a cell able to exist in these two metabolic states may be even more profound. An obvious problem created by extended survival in the absence of exogenous nutrients is the need for a balanced utilization of the available endogenous materials. No essential cell component may reach critically low levels if viability is to be maintained. (Perhaps the "solution" of this problem during evolution was a necessary beginning of what later became the delicate balance of using endogenous materials for both survival and differentiation: "catabolic competition") [53]. Among bacteria, a very interesting phenomenon exists called "substrate-accelerated death". Cells are first grown and subsequently starved in buffer. Addition of the growth-limiting nutrient to the starving cells increases the death rate [37,

38]. The balanced "minimal" metabolism of starvation may be imbalanced to the point of death by forcing the cells to metabolize an added carbon source and thus deplete themselves (critically) of some essential cell component (e.g., ATP, oxidized coenzymes, nitrogen sources, etc.). In this connection, it may be significant that in starving *Streptococcus lactis* the addition of glucose or arginine accelerated the death rate, while the further addition of Mg^{2+} either reduced or eliminated this effect [46]. Apparently the metabolism of cell maintenance dependent upon endogenous substrates is in a delicate balance; when this is upset, death may result from an inadequate resynthesis of critical enzymes, cell membrane or wall material, insufficient osmotic regulation, or a combination of effects such as these. In the evolution of primitive differentiating systems, the acquisition of new permeability barriers concomitant with the initiation of starvation could therefore have selective value by protecting the system from substrate-induced death.

Thus we may have hints of the origin of an apparent incompatibility between growth and differentiation in primitive systems. Of necessity, a system was selected with two metabolically opposed characteristics: (1) the potential to grow rapidly in the presence of exogenous nutrients and (2) the ability to starve in a balanced and efficient manner in their absence. The steps through which differentiating cells evolved could have necessitated a conflict between the metabolism of cell division and differentiation. A sacrifice of characteristics resulting in optimal growth rate might be required in the selection of a cell having characteristics conducive to its survival during starvation. The ability to utilize extensively and efficiently endogenous macromolecules during starvation may not be compatible with a maximum rate of cell multiplication during growth: The latter would require conservation of such material in order to make the optimal number of cells at the expense of an exogenous energy source. Thus, selection for characteristics such as decreased permeability, which would inhibit growth, would also protect and maintain the balanced and precarious metabolism of differentiation.

Starvation, survival and morphogenesis

The next step in evolution may have started with a cell type able to survive for long periods under starvation conditions at the expense of its ability to grow rapidly. The "second conflict" in the evolution of differentiation could result from the compromise inherent in selecting cells able to use endogenous materials both for survival and for the synthesis of a cell wall. Thus, a further demand for metabolic versatility was placed on the system: It must not only be able to alterantely grow and starve, but also to differentiate as it starves. There are many primitive examples of unicellular organisms which accumulate different materials when growth is prevented, for example, by the absence of a nitrogen source. Glycogen, of varying degrees of solubility, accumulates in many organisms [22, 48]; uridine diphosphoglucose as well as glycogen, accumulates in *Agrobacter tumefaciens* [29], trehalose in yeast [35], etc. These processes are reversible in that the soluble materials accumulated rapidly disappear when growth resumes. A review is available on carbohydrate accumulation in the protist as a biochemical model of differentiation [36]. A particularly interesting example of this type has been described in *Streptococcus faecalis* [42]. The omission of an amino acid essential to growth or the inhibition of protein synthesis results in extensive cell wall thickening. Such cells are generally more resistent to deleterious agents, including a cell-wall lytic enzyme from the same organism [43]. Thus the acquisition of a thicker cell wall upon the initiation of starvation (a primitive form of differentiation) may have selective value by protecting the system from adverse environmental conditions (including, perhaps, substrate-accelerated death). Permeability barriers preventing growth must also have been selected when the advantages of completing differentiation outweighed those of simple proliferation. When the substance accumulated under the stimulus of starvation was as insoluble and structured as a spore coat, the process became unidirectional. Under these conditions another mechanism (germination) evolved in order to initiate again the growth phase of the life cycle. Sporulating bacteria and cellular slime molds are examples of systems which alternate a growth and differentiation phase in their life cycle.

Table 16

THE CONSEQUENCES OF AN ISOLATED SYSTEM

To the evolution of differentiation	*To the reproducibility of differentiation*
1. Physical protection (desiccation, heat, toxic materials)	1–4
2. Metabolic protection (substrate-death; loss of endogenous materials)	5. "Programmed starvation" defines overall time period of differentiation; the nature, amount of intermediates and end products
3. High macromolecular turnover, low growth rate	6. At the level of mechanism: initiates and terminates critical pathways, controls the pattern of appearance of intermediates and end products
4. Tendency to alternate growth and differentiation	7. Multiple limiting factors confer stability; the system has "damping" characteristics

In conclusion, some of the possible consequences of isolation to the evolution and reproducibility of differentiation in primitive systems are summarized in Table 14. Perhaps the dependence of differentiation on endogenous metabolism was a necessity, both from the point of view of its evolution and from the point of view of its initiation, control, and termination. The fact of evolution and evidence from comparative biochemistry suggest that our understanding of endogenous substrate control in complex differentiating systems may be furthered through a consideration of its evolution in primitive forms.

REFERENCES

1. ABBOTT, J. and HOLTZER, H. (1966). *J. Cell Biol.* **28**, 473.

2. ADAMSON, D. (1962). *Canad. J. Bot.* **40**, 719.

3. ASTHANA, R. P. and HAWKER, L. E. (1936). *Ann. Botany* (London) **50**, 325.

4. BERNLOHR, R. W. (1965). In *Spores III* (L. L. Campbell and H. O. Halvorson, eds.). Ann Arbor, Mich.: American Society for Microbiology. p. 75.

5. BERG, W. E. (1950). *Biol. Bull.* **98**, 128.

6. BARTH, L. G. and BARTH, L. J. (1951). *J. Exptl. Zool.* **116**, 99.

7. CAHN, R. D. and LASHER, R. (1967). *Proc. Natl. Acad. Sci. U. S.* **58**, 1131.

8. CHALOUPKA, J. and KŘEČKOVÁ, P. (1962). *Biochem. Biophys. Res. Commun.* **8**, 120.

9. DAGGS, R. G. and HALCRO-WARDLAW, H. S. (1933). *J. Gen. Physiol.* **17**, 303.

10. DAWES, E. A. and LARGE, P. J. (1968). *The Endogenous Metabolism of Anaerobic Bacteria.* European Research Office, U.S. Army, Contract No. DAJA37-67-E-0567 University of Hull, Hull, England.

11. DAWES, E. A. and RIBBONS, D. W. (1964). *Bact. Rev.* **28**, 126.

12. DWORKIN, M. (1963). *J. Bact.* **86**, 67.

13. EBERT, J. D. (1965). *Interacting Systems in Development.* New York: Holt, Rinehart, and Winston, Inc.

14. ELSON, D. (1958). *Biochim. et Biophys. Acta* **27**, 216.

15. EPEL, D., WEAVER, A. M., MUCHMORE, A. V., and SCHIMKE, R. T. (1968). *Science* **163**, 294.

16. GALSTON, A. W. (1964). *The Life of the Green Plant.* Englewood Cliffs, New Jersey: Prentice-Hall, Inc.

17. GIBBONS, R. J. (1964). *J. Bact.* **87**, 1512.

18. GRONLUND, A. F. and CAMPBELL, J. J. R. (1965). *J. Bact.* **90**, 1.

19. HABER, A. H., CARRIER, W. L., and FOARD, D. E. (1961). *Am. J. Bot.* **48,** 431.

20. HARRIS, D. L. (1946). *J. Biol. Chem.* **165,** 541.

21. HARRISON, JR., A. P. and LAWRENCE, F. R. (1963). *J. Bact.* **85,** 742.

22. HOLME, T. (1958). *III. Acta Chem. Scand.* **12,** 1564.

23. HUANG, R. C. and BONNER, J. (1965). *Proc. Natl. Acad. Sci. U. S.* **54,** 960.

24. KARLSON, P. and SEKARIS, C. E. (1965). In *Comparative Biochemistry* **6,** (*M. Florkin and H. S. Mason,* eds.) New York: Academic Press Inc., p. 221.

25. KATES, J. R. and JONES, R. F. (1964). *J. Cell. Comp. Physiol.* **63,** 157.

26. LEVISOHN, S. and ARONSON, A. I. (1967). *J. Bact.* **93,** 1023.

27. LUNDBLAD, G. (1949). *Nature* (London) **163,** 643.

28. LUNDBLAD, G. and JOHANSSON, B. (1968). *Enzymologia* **35,** 345.

29. MADSEN, N. B. (1963). *Canad. J. Biochem Physiol.* **41,** 561.

30. MAUPAS, E. (1888). *Arch. Zool. Exper.* **6,** 165.

31. MAYER, A. M. and SHAIN, Y. (1968). *Science* **162,** 1283.

32. MONROY, A., MAGGIO, R., and RINALDI, A. M. (1965). *Proc. Nat. Acad. Sci. U. S.* **54,** 107.

33. NEFF, R. J. and NEFF, R. H. (1969). In *S.E.B. Symposia XXIII,* Dormancy and Survival. New York: Academic Press Inc., p. 51.

34. OPARIN, A. I. (1961). *Life, Its Nature, Origin and Development.* London: Oliver and Boyd, p. 151.

35. PANEK, A. (1962). *Arch. Biochem. Biophys.* **98,** 349.

36. PANNBACKER, R. G. and WRIGHT, B. E. (1967). In, *Chemical Zoology* **1,** Ch. 12 (M. Florkin and B. T. Schaer eds.). New York: Academic Press Inc., p. 573.

37. POSTGATE, J. R. (1967). *Adv. Micro. Physiol.* **1,** 1.

38. POSTGATE, J. R. and HUNTER, J. R. (1964). *J. Gen. Microbiol.* **34,** 459.

39. RAPER, K. B. (1940). *J. Elisha Mitchell Sci. Soc.* **56,** 241–282.

40. SAGER, R. and GRANICK, S. (1954). *J. Gen. Physiol.* **37,** 729.

41. SATO, G. H. and BUONASSISI, V. (1964). In *Metabolic Control Mechanisms in Animal Cells* **13: 1.** (*W. J. Rutter,* ed.). Washington, D.C.: Natl. Cancer Inst. Mon., Govt. Printing Office, p. 81.

42. SHOCKMAN, G. D. (1965). *Bacteriol. Rev.* **29,** 345.

43. SHOCKMAN, G. D. and CHENEY, M. C. (1969). *J. Bact.* **98,** 1199.

44. SPIRIN, A. S. (1966). In *Current Topics in Developmental Biology* **1.** (A. A. Mascona and A. Monroy, eds.). New York: Academic Press, Inc., p. 2.

45. STRANGE, R. E., DARK, F. A., and NESS, A. G. (1961). *J. Gen. Microbiol.* **25,** 61.

46. THOMAS, T. D. and BATT, R. D. (1969). *J. Gen. Microbiol.* **58,** 371.

47. TORREY, J. G. (1966). In *Advances in Morphogenesis* 5. New York: Academic Press Inc., p. 39.

48. TREVELYAN, W. E. and HARRISON, J. S. (1956). *Biochem. J.* **63,** 23.

49. VAHLE, C. (1909). *Zentr. Bakteriol. Parasitenk. Abt. II* **25,** 178.

50. VAN HOUTE, J. and JANSEN, H. M. (1969). *J. Bact.* **101,** 1083.

51. WESTERGAARD, M. and MITCHELL, H. (1947). *Amer. J. Bot.* **34,** 573.

52. WRIGHT, B. E. (1964). In *Comparative Biochemistry* 6 (*M. Florkin and H. S. Mason*, eds.). New York: Academic Press Inc., p. 1.

53. WRIGHT, B. E. (1966). *Science* **153,** 830.

54. WRIGHT, B. E. and Anderson, M. L. (1960). *Biochim. et Biophys. Acta* **43,** 67.

Index

105